AF769488

European School of Oncology

Monographs

Series Editor: U.Veronesi

B. Winograd M. J. Peckham H. M. Pinedo (Eds.)

Human Tumour Xenografts in Anticancer Drug Development

Springer-Verlag Berlin Heidelberg New York
London Paris Tokyo

Dr. Benjamin Winograd
EORTC New Drug Development Office
Free University Hospital
De Boelelaan 1117
NL-1081 HV Amsterdam

Professor Dr. Michael Peckham
British Postgraduate Medical Federation
Central Office
33, Millman Street
London WC1N 3EG/UK

Professor Dr. Herbert Michael Pinedo
Department of Oncology
Free University Hospital
De Boelelaan 1117
NL-1081 HV Amsterdam

The European School of Oncology gratefully acknowledges sponsorship for the Seminar on Human Tumour Xenografts received from the E.O.R.T.C. Early Clinical Trials Group.

Library of Congress Cataloging-in-Publication Data. Seminar on Human Tumour Xenografts (1986 : Milan, Italy) Human tumour xenografts in anticancer drug development. (Monographs / European School of Oncology) "The European School of Oncology gratefully acknowledges sponsorship for the Seminar on Human Tumour Xenografts received from the E.O.R.T.C. Early Clinical Trials Group" - T.p. verso. Seminar held 26th-27th May 1986 in Milan. Includes bibliographies and index. 1. Antineoplastic agents - Testing - Congresses. 2. Xenografts - Congresses. 3. Tumors - Transplantation - Congresses. 4. Cancer - Animal models - Congresses. I. Winograd, B. (Benjamin), 1955 - . II. Peckham, M.J. (Michael J.) III. Pinedo, H.M. IV. European School of Oncology. V. E.O.R.T.C. Early Clinical Trials Group. VI. Title. VII. Series: Monographs (European School of Oncology) [DNLM: 1. Antineoplastic Agents - therapeutic use - congresses. 2. Drug Screening - congresses. 3. Neoplasms - drug therapy - congresses. 4. Transplantation, Heterologous - congresses. QZ 267 S471h 1986] RC271.C5S48 1986 616.99'400724 88'2034

ISBN-13: 978-3-642-73254-6 e-ISBN-13: 978-3-642-73252-2
DOI: 10.1007/ 978-3-642-73252-2

Softcover reprint of the hardcover 1st edition 1988

Typesetting (media conversion)
2123/3145-543210

Foreword

Starting in 1986, the European School of Oncology has expanded its activities in postgraduate teaching, which consisted mainly of traditional disease-orientated courses, by promoting new educational initiatives. One of these is the cloister seminars, short meetings intended for highly qualified oncologists and dealing with specific, controversial aspects of clinical practice and research. Another is the institution of permanent study groups, also called task forces, where a limited number of leading experts are invited to meet once a year with the aim of defining the state of the art and possibly reaching a consensus on developments and treatment in specific fields of oncology.
This series of ESO Monographs was designed with the specific purpose of disseminating the results of the most interesting of the seminars and study groups, and providing concise and updated reviews of the subjects discussed.
It was decided to keep the layout very simple in order to keep costs to a minimum and make the monographs available in the shortest possible time, thus overcoming a common problem in medical literature: that of the material being outdated even before publication.

Umberto Veronesi
Chairman, Scientific Committee
European School of Oncology

Position Paper on the Application of Human Tumour Xenografts as a Model for Preclinical Phase II Studies in the Evaluation of New Anticancer Compounds

Outcome of a seminar on Human Tumour Xenografts organised by the European School of Oncology, Milan, 26th–27th May, 1986.

Nude mice and artificial immune-deficient mice are unique hosts for human tumours in order to study their biology and to permit treatment in an *in vivo* situation. During the seminar a consensus was reached between the major European Institutes as to how this model system could be used in the evaluation of anti-cancer therapies.

A good correlation between anti-tumour responses observed in the nude mouse and in man has been documented in those situations where the same tumour could be established as a xenograft and treated with the same anti-cancer agents in the patient. The time to grow and for an anti-tumour response to be documented precludes the use of this system as a predictive test of drug sensitivity in individual patients.

Several of the participating centers are focusing their work on 1 or 2 tumour types. The results of these studies show that a number (5–15) of different tumour lines can identify clinically active agents for that particular tumour type. Responses observed in the xenografts are, with a few exceptions (e.g. methotrexate and 5-fluorouracil), comparable to those observed clinically for single agents.

At present no pre-clinical tests are available which predict the anti-tumour activity of new compounds in specific tumour types. For this reason, numerous phase II clinical studies have to be performed and as a result large numbers of patients are treated without therapeutic benefit in order to detect anti-tumour activity in each tumour type. Furthermore, only a few active anti-tumour compounds have entered the clinic as a result of such studies during the past few years. To illustrate the problem the Early Clinical Trials Group of the EORTC has studied more than 20 new compounds during the past 12 years in single-agent phase II studies in a range of solid tumour types. More than 2500 patients have been entered in these studies but the response rates have been extremely disappointing.

The seminar concluded with a proposal to carry out pre-clinical phase II studies in the xenograft model prior to evaluating new anti-cancer agents in clinical phase II studies. In our view this can only be achieved if a collaboration can be established based on the experience and expertise existing in the major European institutes. The tumour lines of the institutes participating in the seminar will be made available and a mechanism will need to be established to ensure close feedback between participating institutes and the laboratory responsible for carrying out the pre-clinical phase II studies.

B. Winograd
M. J. Peckham
H. M. Pinedo

Introduction

This Monograph gives an overview of papers presented at the Seminar entitled "Human Tumour Xenografts in Anticancer Drug Development", organized by the European School of Oncology, May 25–26, 1986.
Since 1969, when the first successful heterotransplantation of a human malignant tumour in the "nude" mouse was reported, this model has been used in various experimental studies on human tumour morphology, biology, biochemistry and antitumour therapy. Due to the fact that each individual tumour largely retains its characteristics through serial transplantation in the immune-deficient mouse, this model seems valuable in the development of anticancer drugs.
Conclusions of the presented papers and discussions during the Seminar resulted in the position paper on the application of human tumour xenografts as a model for preclinical phase II studies, presented hereafter.
A research group of several European investigators was created and the feasibility and predictability of preclinical phase II studies is at present being evaluated in a multicenter study. If this approach will prove feasible and a high predictive value for active and inactive new anticancer compounds can be established, the number of clinical phase II studies without therapeutic benefit for the patients will be reduced and an increased number of new anticancer treatments and strategies can be evaluated in the clinic.

Table of Contents

VIII. Importance of Drug Resistance in Screening

IX. Proferred Papers at the E. S. O. Seminar

List of Contributors

Aamdal, S. 63, 136
Aapro, M.S. 133
Abbott, B.J. 115
Anderson, L.E. 135
Arione, R. 139
Baldwin, R.W. 95
Balkwill, F.R. 91
Bieri, A. 135
Blythman, H.E. 103
Bord, A. 103
Boven, E. 11, 33, 71
Boyd, M.R. 115
Braakhuis, B.J.M. 37, 81
Buisson, I. 103
Canevari, S. 141
Carmichael, J. 51
Chatel, M. 134
Chiodetti, N. 135
Clegg, J.A. 95
Colnaghi, M.I. 141
Courtney, E.R. 127
Cunningham, D. 127
Darcel, F. 134
Engelholm, S.A. 141
Fergusson, R.J. 99
Fiebig, H.H. 25, 137
Fine, D.L. 115
Floersheim, G.L. 135
Fodstad, Ø. 15, 63, 115, 136
Forni, G. 139
Fox, B.W. 123
Freshney, R.I. 127
Garnett, M.C. 95
Gasteren, van J.J.M. 137
Giuliani, F.C. 75
Gorelik, E. 115
Hamilton, T. 127
Hinkelbein, W. 137
Hoogenhout, J. 137
Jansen, F.K. 103
Jerusalem, C.R. 137
Kal, H.B. 137
Kaye, S.B. 127
Kelland, L.R. 57
Koepf-Maier, P. 85
Kopper, L. 138
Lapis, K. 138
Leyva, A. 81
Lobbezoo, M.W. 111
Martinotti, M.G. 139
Mattern, J. 47
Mayo, J.G. 115
McLemore, T.L. 115
Menard, S. 141
Merry, S. 127
Mezzanzanica, D. 141
Moulinoux, J.Ph. 134
Nagy, P. 138
Orlandi, R. 141
Osieka, R. 41
Parmiani, G. 140
Peckham, M.J. 3
Pezzoni, G. 75
Pihl, A. 63, 136
Pimm, M.V. 95
Pinedo, H.M. 81, 111
Pratesi, G. 140
Quemener, V. 134
Rajnay, J. 138
Richer, G. 103
Ripamonti, M. 141
Shoemaker, R.H. 115, 136
Smyth, J.F. 51, 99, 135
Snow, G.B. 37, 81
Spany-Thomsen, M. 141
Steel, G.G. 3
Tagliabue, E. 141
Thurneyssen, O. 103
Tonkin, K.S. 57
Tortoreto, M. 140
Volm, M. 47
Wallbrecher, E. 137
Wayss, K. 47
Wennerberg, J. 142
Winograd, B. 111
Wuerdinger, A. 137

I. Biological Behaviour of Human Tumours Transplanted in Mice and the Evaluation of Therapeutic Response

The Therapeutic Response of a Variety of Human Tumour Xenografts

G. G. Steel and M. J. Peckham

Radiotherapy Research Unit, Institute of Cancer Research, Clifton Avenue, Sutton, Surrey SM2 5PX, United Kingdom

This paper reviews a research programme on human tumour xenografts that has been under way since around 1975. Its objectives have been to establish xenografts that are representative of a number of tumour types, to investigate their biological characteristics and to evaluate their therapeutic response. We have been interested in the evaluation of response both to chemotherapy and radiotherapy. During the period 1975–1983 our major effort was on response to cytotoxic drugs. At that time it was recognised that chemotherapeutic agents are more selective than radiation in killing tumour cells, for instance, the range of sensitivities observed for different tumour cell types treated with a particular drug is much wider than for radiation (26). We reasoned that since human tumours no doubt change their biological characteristics to some degree during transplantation from man to mouse, it was more likely that the ranking of chemotherapeutic response would be maintained than that response to radiation would do so, especially since *in vivo* radiation response is so dependent on conditions of local oxygenation. Although this reasoning is probably still correct, we have more recently appreciated the importance of radiobiological studies directly on human tumour cells and our research emphasis has therefore swung to radiobiology. This research programme has largely been carried out by a succession of clinical scientists (see Acknowledgements) and a number of earlier reviews have appeared (27, 29, 30).

Immune-Suppressed Mice as Hosts for Xenografts

The distinctive characteristic of this programme has been the use of immune-suppressed conventional mice, as opposed to congenitally immune-deficient animals. The choice of hosts and immune-suppression techniques was reviewed in 1982 (29). It would seem that the most successful approach is to treat mice with a combination of neonatal thymectomy and whole-body irradiation to a dose of about 9 Gy. This is a supralethal radiation dose and it is necessary to protect the haemopoietic system if the animals are to survive. Giving unirradiated syngeneic bone marrow immediately after whole-body irradiation is a successful procedure both in terms of mouse survival and xenograft receptivity, especially if the size of the marrow graft is kept as low as possible. A perhaps more effective procedure is to use cytosine arabinoside (Ara-C) as a protective agent. The phenomenon thus being exploited is a remarkable one which has been the subject of a considerable amount of research by my colleagues in this Institute (17, 20). Ara-C must be given before irradiation, and the timing is fairly critical: 48 h in our experience. There is evidence that this has the effect of stimulating the bone marrow to repopulate rapidly after irradiation: intestinal epithelium is also protected, though the optimum effect in this tissue occurs with only a 12-hour gap between Ara-C and irradiation (20).

The main disadvantage of immune-suppressed (as opposed to nude) mice is their tendency to recover the ability to reject human cells during the few months following irradiation. We have found that the duration of the period of high receptivity can be increased by raising the radiation dose to 10 or even 11 Gy (31). However, at these very high doses, the mice are more fragile and require support with antibiotics. A range of additional procedures have been used in an effort to improve the take rate of breast carcinoma xenografts but they were unsuccessful (1).

Our experience in establishing new xenograft lines from a range of human tumours is summarised in Table 1. The main emphasis has been on colorectal tumours (52% success rate), breast carcinomas (10%), bronchial carcinomas (78%), testicular teratomas (30%) and ovarian carcinomas (25%). These success rates indicate the proportion of clinical specimens that yielded serially transplantable lines.

Each of these tumour types was studied by a different scientist and the technique of immune deprivation has varied over the period of this work. The most recent studies, on cervix and bladder tumours and neuroblastomas, have employed nude mice. It would, however, seem clear that breast and ovarian tumours (especially ascites) have generally been more difficult to graft, with melanomas and tumours of the colon and bronchus being relatively successful.

Table 1. Summary of xenograft success rates[a]

Tumour location and type		Number of patients tested	Number of lines established	Success rate %	Refs.
Colon-rectum		65	34	52	18, 21
Stomach		1	0		21
Pancreas		2	1	50	21
Glioma		5	0		21
Melanoma		15	9	60	21
Breast		80	8	10	1
Bronchus:	adenocarcinoma	6	6	100	25
	squamous	17	11	65	25
	small-cell	18	15	83	25
	large-cell	8	6	75	25
Testicular tumours:	MTU[b]	26	9	35	14
	MTI	17	3	18	14
	MTT	6	2	30	14
	seminoma	12	0		14
	mixed	8	3	38	14
Ovary:	solid	44	11	25	15
	ascites	9	0		15
Cervix:		22	9	41	16
Bladder:		12	3	25	19
Neuroblastoma:		14	3	21	9, 10
		387	133	34	

[a] In immune-suppressed conventional mice, except in the case of the last three tumour categories, where nude mice were used.
[b] MTU = embryonal, MTI = teratocarcinoma, MTT = choriocarcinoma.

The maintenance of biological characteristics in xenografts has been documented in a number of ways. Karyotype analysis has frequently been performed and with one exception the predominant karyotype was human. Although the occasional mouse karyotype was noted, there was one case of a rapidly growing tumour which appeared at the implantation site of a human teratoma xenograft and which was found to have an exclusively mouse karyotype. Histological appearance of the source tumours has also been well maintained, and the documentation of ectopic hormones in bronchial carcinoma xenografts (24), the production of alphafetoprotein by teratona xenografts and carcinoembryonic antigen by colon carcinoma xenografts also supports the view that xenografts do maintain some of the biological characteristics of the source tumour.

Our experience of the growth rate of human tumour xenografts (30) is that in the initial passage from man to mouse, the mean volume doubling times for various disease categories have ranged from 14 to 30 days. Growth acceleration during serial passage is common and by the fourth passage, the corresponding range is 6 to 15 days. It is our view that these growth rates are considerably higher than for measurable metastases in man and that this probably reflects the failure of some slowly-growing human tumours to transplant successfully, as well as a tendency for cell selection during passage to favour rapidly growing cells.

The In Situ Response of Xenografts to Chemotherapy

Much of the work in the present programme has employed tumour growth delay as the end-point of effect. A group of immune-suppressed mice is implanted with tumours, usually within a few days of the immune-suppressive radiation treatment. Not all implants grow at the same rate or with the same latency period. When a good number of them have grown beyond about 7 mm diameter the tumours are accurately measured. Those falling within a defined range of size (say 8–10 mm average superficial diameter) are selected and split into comparable groups, each containing 6–10 tumours. One group is kept as a control, the others are treated with one or more drugs at one or more dose levels. Caliper measurements are then continued approx. 3 times per week and tumour volumes calculated either by a geometrical formula or preferably using a calibration curve technique (26). Growth delay is then measured by determining the time (T1) taken for control tumours to double in volume, the time (T2) for treated tumours to double, and then calculating

$$\text{Specific growth delay} = \frac{T2 - T1}{T1} \qquad (1)$$

This parameter can be regarded as an estimate of the number of volume doubling times by which growth is delayed; it may therefore provide a basis for comparison of therapeutic response between tumours of different growth rates.

The choice of drug dose is another difficult problem in xenograft chemotherapy studies, especially where the intention is to compare the effectiveness of a series of drugs. The procedure used throughout the present group of studies has been to employ each drug at its maximum tolerated dose in mice. The rationale behind this lies in the claim by

Freireich et al. (13) that a linear relationship exists between the LD10 doses of cytotoxic drugs in mice and their maximum tolerated doses in man.

Evidence for Validity of Xenografts In Determining Response to Chemotherapy

Human tumours do maintain on xenografting the broad characteristics of their response to chemotherapy, in spite of changes in their growth rate and other biological features. The evidence for this is of 2 types: comparison of response in groups of xenografts with general clinical experience; secondly, direct comparison in a small group of patients of clinical response and xenograft response.

Our overall experience in studying the response to chemotherapy of 6 types of human tumour xenografts is summarised in Fig. 1. These data are the composite results of a number of research projects (30). In each case the specific growth delay at an LD10 drug dose has been calculated. The actual drug doses used varied slightly from one study to another and the teratomas and small-cell lung tumours were so sensitive to treatment that in these cases it was necessary to stay well below the LD10 level, in order to avoid cures. The growth delay values have therefore been adjusted to the LD10 dose by linear extrapolation or interpolation (see ref. 30 for details). The choice of drugs used against each tumour was influenced by their known clinical activity or potential interest. The panels of data in Fig. 1 are arranged in order of decreasing tumour re-

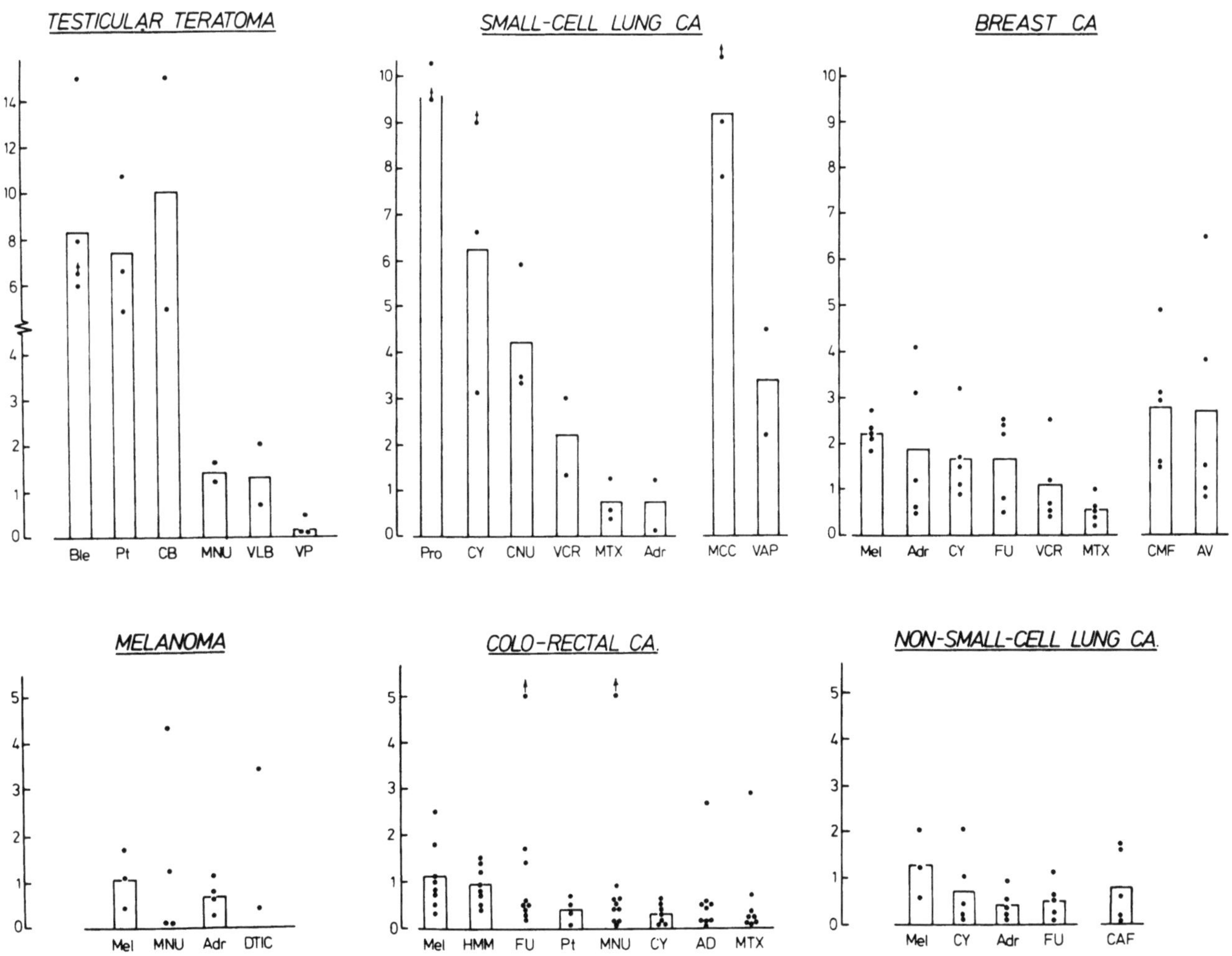

Fig. 1. The comparative response of six types of human tumour xenografts to chemotherapy. In each panel, the ordinate indicates the specific growth delay that resulted from a maximum tolerated dose of chemotherapeutic drug. Each point indicates the response of a particular patient's xenografts to the drug in question. BLE = bleomycin; Pt = cisplatin; CB = carboplatin; MNU = MeCCNU; VLB = vinblastine; VP = VP16, Pro = procarbazine; CY = cyclophosphamide; CNU = CCNU; VCR = vincristine; MTX = methotrexate; Adr = doxorubicin; Mel = melphalan; FU = 5-fluorouracil; HMM = hexamethylmelamine; AD = actinomycin D. Drug combinations (also at maximum tolerated dose): MCC = MTX + CY + CNU; VAP = VCR + Adr + Pro; CMF = CY + MTX + FU; AV = Adr + VCR; CAF = CY + Adr + FU

sponse. Colo-rectal carcinomas, non-small-cell lung tumours and melanomas did not significantly exceed an average specific growth delay of 1.0 with any drug. Breast carcinomas responded well to 4 of the single agents as well as to the drug combination (2). Small-cell lung tumours were cured by large doses of cyclophosphamide, procarbazine, or the MCC combination (Methotrexate, cyclophosphamide, CCNU). The testicular tumours were similarly very responsive: the growth delay values given have in most cases been extrapolated from results actually obtained with doses around one-fifth of the LD10 level. These data broadly reflect clinical experience with chemotherapy in these disease areas.

Comparison of Xenografts with Donor Patient Response

The most precise and relevant way of validating the xenograft system for therapeutic purposes is directly to compare xenograft response with the clinical response of the donor patient. This is logistically a difficult type of investigation because of the combined effect of unavoidable factors: the proportion of donor patients who receive chemotherapy may be small; few of them may give an objectively measurable clinical response; the xenograft success rate may be below 50%. In spite of these limitations, comparisons of this type have been possible in 3 disease areas. Four of the colorectal cancer patients studied by Nowak et al. (18) received chemotherapy (MeCCNU, 5-FU and DTIC); in 3 there was no evidence of a therapeutic response and the disease progressed; one achieved an objective clinical response, and the xenografts from this patient were the most responsive (of the 10 xenograft lines studied) to MeCCNU and also responsive to 5-FU. Two of the donor patients in the melanoma study of Selby et al. (22) had measurable lung metastases and were treated with melphalan. One (donor of HX47) showed an objective response but the other (HX41) did not. The melphalan cell survival curve for the xenografts was twice as steep in HX47 as in HX41. Our major study of this type was that of Shorthouse et al. (25) using bronchial carcinomas. Xenograft and donor patient response was evaluated in 7 small-cell carcinomas, 3 squamous carcinomas, 4 adenocarcinomas and 1 large-cell carcinoma. The clinical response of small-cell tumours was reflected in highly responsive xenografts, as was the lack of response of the non-small-cell tumours. Two previously-treated patients with small-cell cancer failed to achieve a clinical response, and the xenografts were similarly resistant.

The consistency of agreement between patient and xenograft response in these studies is encouraging. This does of course not mean that there is a realistic prospect of using xenografts for clinical chemosensitivity testing, since, as has been discussed (3), the limitations of speed of assay and of success rate would limit the proportion of patients helped to only 10–20%. However, the use of xenografts as models of particular human cancers is supported.

Cell Survival Studies In Xenografts

The principal alternative to *in situ* methods of measuring therapeutic response in xenografts is to determine clonogenic cell survival. Tumour-bearing mice are treated, the tumours are excised within about a day, disaggregated into a single-cell suspension and the cells placed in an *in vitro* (or perhaps *in vivo*) environment in which their ability to form a colony can be evaluated. We have used a number of clonogenic assays for xenografted tumours, including *in vitro* assay in soft agar (6) or in monolayer (16), agar diffusion chamber assay (23) and a lung colony assay (34). *In vitro* plating efficiencies vary widely from very low levels in some tumours to around 30% in favourable cases. Problems of tissue disaggregation are sometimes severe and at the present time only a minority of xenografted tumours are amenable to reliable clonogenic assay. A recent study of squamous carcinomas of the cervix (16) found that the plating efficiency in soft agar was very low. However, a colony assay in monolayer over a feeder layer of 3T3 fibroblasts was developed and plating efficiencies of 2–30% have been achieved in 4 tumour lines.

The great advantage of clonogenic cell survival as an end-point for xenograft studies is that by removing the treated and control cells into a uniform environment, problems of host defence mechanisms are avoided. Other advantages of this approach are its sensitivity (often down to a surviving fraction of 10^{-3}) and its ability to detect the presence or absence of resistant components within the tumour or a 'shoulder' at low drug doses.

Chemotherapy studies of cell survival in the present research programme have included studies on pancreatic and bronchial carcinomas, melanomas, and a single survival curve for a colon carcinoma. Survival curves were in almost all cases indistinguishable from an exponential, with rare examples of a small shoulder. The only marked exception to this was in the response of 2 of the melanomas to DTIC, where the curves fell to a clear plateau at about 10^{-2} survival for drug doses in excess of 100 mg/kg.

A comparative study (15) of 4 ovarian carcinoma xenografts in which the tumour cells were exposed to melphalan, cisplatin and carboplatin *in vitro* before assessment of clonogenic cell survival found fairly good agreement among the tumours. The main objective was to compare carboplatin with cisplatin in these tumours and the ratio of doses to give the same effect (carboplatin/cisplatin) ranged from 7 to 16.

Induced Drug Resistance In Xenografts

The induction of drug resistance in small-cell bronchial carcinoma xenografts was studied by Dr. R. Berman and his results are described in 2 publications (4, 5). He chose 3 xenograft lines established by Shorthouse (25), HX72, HX78 and HX88. The donor of HX72 had received chemotherapy prior to xenografting and there was clinical evidence for drug resistance. It was found that in terms of xenograft growth delay HX72 cells were less sensitive to cyclophosphamide than the other 2 lines. Efforts to induce drug resistance by repeated treatment of the xenografts, once in each passage, were only partially successful. Only the HX78 line developed resistance; the specific growth delay for a fixed cyclophosphamide dose fell to half after 3 treatments and to a quarter after 6 treatments. Subsequent studies of this induced-resistant line (HX78CY) showed that it was cross-resistant to melphalan, MeCCNU, vincristine and to X-radiation. Growth rate was unchanged, as was the general histological appearance of small-cell lung cancer. However, staining with antisera against keratin demonstrated the appearance of small amounts of keratin in HX78CY. It was concluded that the repeated cyclophosphamide treatment had led to the development of a degree of squamous differentiation in the xenografts, probably by allowing pre-existing squamous elements to become more prominent. The observed features of cross-resistance would be consistent with this.

Response to Radiation of Xenografted Human Tumour Cells

Since 1982 the main emphasis of our research on xenografts has turned to studies of radiation sensitivity. Following the earlier observations by Fertil & Malaise (12) a review of radiosensitivity in human tumour cell lines irradiation *in vitro* led to the conclusion that the initial slope of the cell survival curve (quantified by the surviving fraction at 2 Gy) correlated with the clinical response of tumours to radiotherapy (8). Thus, although it may still be true as suggested at the beginning of this article that in situ studies of the radiobiology of xenografts are influenced by a somewhat artificial level of hypoxia, studies of the *in vitro* sensitivity could be very important. And xenografts provide a valuable means of establishing and maintaining human tumours in the laboratory for this purpose. The ground-work for these studies was carried out by V. D. Courtenay

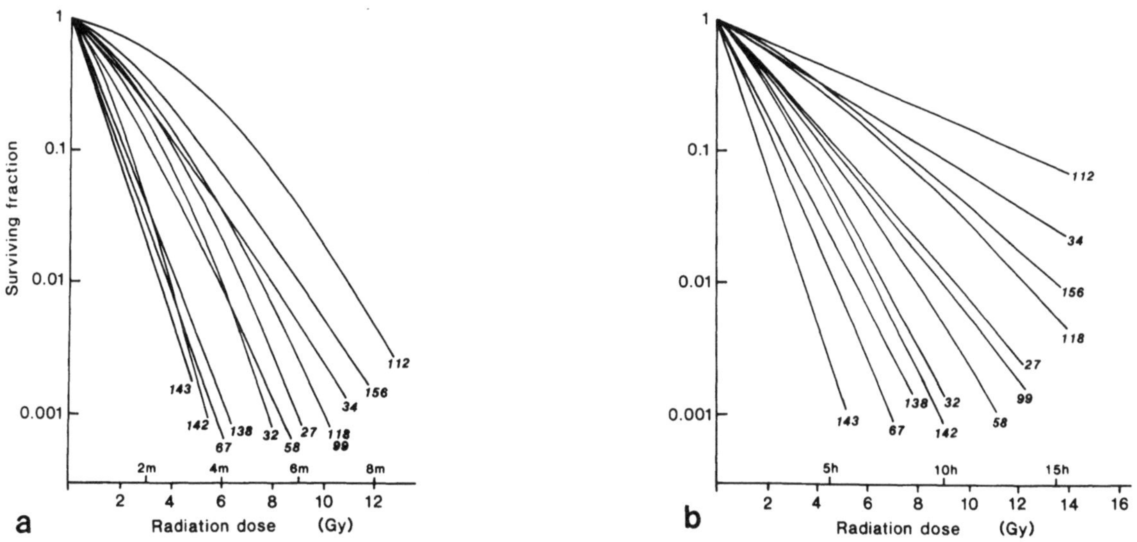

Fig. 2a, b. Cell survival curves for 12 human tumour cell lines irradiated (**a**) at high dose rate (approx. 150 cGy/min) or (**b**) at low dose rate (approx. 1.6 cGy/min). The duration of exposure has been marked on the abscissa. The human tumour lines are indicated by numbers: 138, 142, 143 neuroblastoma; 67, 112 bladder; 32, 58 pancreas; 34, 118 melanoma; 156 cervix; 27 teratoma; 99 breast

who developed the double-layer soft agar cloning technique for human tumour cells (6). This technique was used to investigate radiosensitivity in human tumour cells, especially melanomas. It was found that the cell survival curves for cells taken from xenografts or direct from the patient were the same (28).

Subsequent investigations have demonstrated the relatively high radiosensitivity of neuroblastomas (9, 10) and the implications of the results for management of this disease have been discussed. The radiosensitivity of human lung tumour lines has been studied (11) and significant differences in response between the various tumour sub-types were observed.

In a variety of tumour types, studies have been made of the extent of sparing of radiation damage as the dose-rate is lowered (32, 33). Fig. 2 summarises an extensive amount of work in this area, showing the range of acute radiation cell survival curves found among a variety of human tumour types. When the radiation dose rate is lowered (right-hand panel), repair is allowed to take place during irradiation and cell survival increases. Some of the more radiosensitive cell lines show little sparing but in others the amount of low dose-rate sparing is considerable. It is clear that the extent and perhaps the speed of repair of radiation damage differs among the different tumour types and that this may be the reason for observed differences in radioresponsiveness. Low dose-rate studies therefore provide a useful method of investigating these processes in human tumour cells.

Acknowledgements

The experimental work described here has largely been the work of our colleagues: M.J. Bailey, R. Berman, V.D. Courtenay, J.M. Deacon, G.M. Duchesne, A.C. Jones, L.R. Kelland, K. Nowak, J.H. Peacock, A.Y. Rostom, P.J. Selby, A.J. Shorthouse, and J.M. Thomas.

References

1. Bailey MJ, Gazet JC and Peckham MJ: Human breast cancer xenografts in immune-suppressed mice. Br J Cancer 1980 (42): 524-529
2. Bailey MJ, Gazet JC, Smith IE, Steel GG: Chemotherapy of human breast-carcinoma xenografts. Br J Cancer 1980 (42): 530-536
3. Bailey M, Jones A, Raghavan D, Selby P, Shorthouse A, Gibbs J, Peckham MJ: Limitations of human tumour xenografts in individual patient drug-sensitivity testing (Abstract). Br J Cancer 1981 (43): 725
4. Berman R, Gusterson B and Steel GG: Resistance to alkylating agents and tumour differentiation in xenografts of small cell lung cancer. Br J Cancer 1985 (51): 653-658
5. Berman R and Steel GG: Induced and inherent resistance to alkylating agents in human small-cell bronchial carcinoma xenografts. Br J Cancer 1984 (49): 431-436
6. Courtenay VF, Mills J: An in vitro colony assay for human tumours grown in immune-suppressed mice and treated in vivo with cytotoxic agents. Br J Cancer 1978 (37): 261-268
7. Courtenay VD, Mills J, Steel GG: The spectrum of chemosensitivity of two human pancreatic carcinoma xenografts. Br J Cancer 1982 (46): 436-439
8. Deacon J, Peckham MJ and Steel GG: The radioresponsiveness of human tumours and initial slope of the cell survival curve. Radiother and Oncol 1984 (2): 317-323
9. Deacon JM, Wilson PA and Peckham MJ: The radiobiology of human neuroblastoma. Radiother and Oncol 1985 (3): 201-209
10. Deacon JM, Wilson PA and Steel GG: Radiosensitivity of neuroblastoma. Advances in Neuroblastoma Research 1985: 525-531
11. Duchesne G, Peacock J and Steel GG: The acute in vitro and in vivo radiosensitivity of human lung tumour lines. Radiother and Oncol 1986 (7): 353-361
12. Fertil B and Malaise EP: Inherent radiosensitivity as a basic concept for human tumour radiotherapy. Int J Radiat Oncol Biol Phys 1981 (7): 621-629
13. Freireich EJ, Gehan EA, Rall DP, Schmidt LH, Skipper HE: Quantitative comparison of toxicity of anticancer agents in mouse, rat, hamster, dog, monkey and man. Cancer Chemother Rep 1966 (50): 219-234
14. Gibbs J, Peckham MJ and Steel GG (unpublished)
15. Jones AC, Wilson PA, and Steel GG: Cell survival in four ovarian carcinoma xenografts following in vitro exposure to melphalan, cisplatin and JM8. Cancer Chemother Pharmacol 1984 (13): 109-113
16. Kelland LR, Burgess L, and Steel GG: Characterisation of four new cell lines derived from human squamous carcinomas of the uterine cervix. Cancer Res 1987 (47): 4947-4952
17. Millar JL, Blackett NM, Hudspith BN: Enhanced postirradiation recovery of the haemopoietic system in animals pretreated with a variety of cytotoxic agents. Cel Tiss Kinet 1978 (11): 543-553
18. Nowak K, Peckham MK, Steel GG: Variation in response of xenografts of colo-rectal carcinoma to chemotherapy. Br J Cancer 1978 (37): 576-584
19. Peacock JH (unpublished)
20. Phelps, TA: Cytarabin (Ara-C) induced radioresistance of mouse jejunal stem cells following single or fractionated doses of radiation. Int J Radiat Oncol Biol Phys 1980 (6): 1671-1677
21. Pickard RG, Cobb LM, Steel GG: The growth kinetics of xenografts of human colorectal tumours in immune-deprived mice. Br J Cancer 1975 (31): 36-45
22. Selby PJ, Courtenay VD, McElwain TJ, Peckham MJ and Steel GG: Colony growth and clonogenic cell survival in human melanoma xenografts treated with chemotherapy. Br J Cancer 1980 (42): 438-447
23. Selby PJ and Thomas JM: Clonogenic cell survival curves for human melanoma xenografts using agar diffusion chamber and lung colony assays. Br J Cancer 1980 (41 Suppl IV): 150-153
24. Shorthouse AJ, Carter SM, Ellison ML: Tumour marker production in human bronchial carcinoma xenografts. Oncodevelop Biol Med 1982 (3): 273-281
25. Shorthouse AJ, Jones JM, Steel GG and Peckham MJ: Experimental combination and single-agent chemotherapy in human lung-tumour xenografts. Br J Cancer 1982 (46): 35-44
26. Steel GG: The growth kinetics of tumours. Oxford University Press, London 1977, pp 255-265
27. Steel GG: Therapeutic response of human tumour xenografts in immune-suppressed mice. In: Sordat B (ed) Immune-deficient animals in experimental research. Karger AG, Basel 1984, pp 395-404

28. Steel GG: How well do xenografts maintain the therapeutic characteristics of the source tumour in the donor patient? In: RF Kallman (ed) Rodent tumours in experimental cancer therapy. 1987 pp 205-208
29. Steel GG, Courtenay VD and Peckham MJ: The immune-suppressed mouse as an alternative host for heterotransplantation. In: Fogh J and Giovanella BC (eds) The nude mouse in experimental and clinical research, Vol. 2. Academic Press Inc 1982, pp 207-227
30. Steel GG, Courtenay VD and Peckham MJ: The response to chemotherapy of a variety of human tumour xenografts. Br J Cancer 1983 (47): 1-13
31. Steel GG, Courtenay VD, Phelps TA and Peckham MJ: The therapeutic response of human tumour xenografts. In: Sparrow S (ed) Symposium on immuno-deficient animals in cancer research. McMillan, London 1980, pp 179-189
32. Steel GG, Deacon JM, Duchesne GM, Horwich A, Kelland LR and Peacock JH: The dose-rate effect in human tumour cells. Radiother and Oncol 1987. Radiother and Oncol 1987 (9): 299-310
33. Steel GG, Down JD, Peacock JH and Stephens TC: Dose-rate effects and the repair of radiation damage. Radiother and Oncol 1986 (5): 321-331
34. Thomas JM: A lung colony clonogenic cell assay for human malignant melanoma in immune-suppressed mice and its use to determine chemosensitivity, radiosensitivity and the relationship between tumour size and response to therapy. Br J Surg 1979 (66): 696-700

Biology of Human Ovarian Cancer Xenografts

E. Boven

Department of Oncology, Free University Hospital, De Boelelaan 1117, 1081 HV Amsterdam, The Netherlands

Abstract

In an attempt to establish human ovarian cancer tissue samples as subcutaneous xenografts in athymic (nu/nu) mice, our overall take rate was 32% and 11% of all specimens could be transferred beyond 3 passages. Transplantability was most successful if samples were derived from a metastatic lesion, contained a high number of mitotic figures and were not infiltrated by lymphocytes. The take rate was not dependent on the histologic subtype or the degree of differentiation.

All tumor lines thus far in use for chemotherapy studies have retained the histologic pattern of origin and have a consistent growth rate. At present our human ovarian cancer model contains a series of lines which vary in histologic subtype and tumor doubling time, differences that may also be expected in a group of ovarian cancer patients.

Introduction

Many human solid malignancies can be grown as subcutaneous xenografts in athymic (nu/nu) or artificially immune-deprived mice. While tumor lines of colon cancer, melanoma and soft-tissue sarcoma are relatively easy to establish, other tumor types, such as breast cancer and ovarian cancer, are difficult to grow. It is of interest to investigate whether specific factors can be detected in association with a particular malignancy that influence the growth as a xenograft. Once a tumor line has been established, it generally retains the histologic pattern and biologic characteristics of origin (1-3). Characterization may include a series of parameters of which histology and growth rate have been most often applied.

In order to develop a panel of human ovarian cancer lines reflecting a group of patients, we determined the possible relation between the take rate and several tumor-specific aspects and characterized our tumor lines using various parameters.

Materials and Methods

Animals and Transplantation

Female athymic (nu/nu) mice of the B10 LP/cpb strain (Centraal Proefdieren Bedrijf, Zeist, NL) were maintained in isolation in sterile cages with paper filter covers. Animal handling was done under sterile conditions in a laminar-down-flow hood. Fresh tumor specimens obtained from ovarian cancer patients were processed within 2 to 6 hours. Fragments of 2-3 mm in diameter were implanted subcutaneously in both flanks (behind the front leg) of 6 animals, each 8 to 10 weeks of age. The mice were observed for 6 months for signs of tumor growth. Progressive growth beyond a volume of 50 mm^3 (0.5 × length × width × thickness) in any of the 6 mice was considered as a positive take.

Take Rate Evaluation

A stepwise logistic regression analysis was applied to investigate whether specific factors were of influence on the transplantability of the tumor sample. The origin of the tissue and several data obtained from light microscopy were independent variables (Table 1), whereas the success of transplantation was the dependent variable.

Table 1. Independent variables of the human ovarian cancer tissue sample in relation to the take rate

Variable	Classification
Tumor origin	Primary, metastatic
Histologic subtype	Serous, mucinous, endometrid, clear-cell, undifferentiated
Differentiation	Good, moderate, poor
Mitotic figures (5 HPF)	≤5, >5
Quantity of tumor tissue	<50%, ≥50%
Lymphocytic infiltration	−, +/−, +, ++
Necrotic areas	−, +/−, +, ++

Table 2. Human ovarian cancer lines in nude mice: histologic subtype and tumor doubling time T_D (days)

Tumor line	Histology	T_D
Ov. Pe	moderately differentiated mucinous	18
MRI-H-207	undifferentiated	3.5
Ov. Me	carcinosarcoma	7
Ov. He	moderately differentiated mucinous	13
Ov. G1	poorly differentiated serous	16
Ov. Ri (C)	moderately differentiated serous	14
FCo	moderately differentiated clear cell	17
FKo	moderately differentiated serous	16
FMa	poorly differentiated mucinous	7
A 2780S	undifferentiated	4

Characterization of Tumor Lines

Tumor lines in use for chemotherapy studies (Table 2) have been described previously (4).
These lines were examined for a series of parameters in subsequent passages. Besides the usual histologic staining procedure (haematoxylin and eosin), additional stains included Periodic Acid Schiff (PAS), PAS/diastase, Alcian Blue (AB) and the indirect staining procedure for the detection of carcinoembryonic antigen (CEA). For LDH-isoenzyme identification tumor extracts were prepared and analyzed as described before (4). Chromosomes were made visible with quinacrine mustard (Q-banding) after preparation of a single-cell suspension (4).
The tumor doubling time (T_D) was calculated in the exponential growth phase, which was the period (in days) from 50 mm^3 to 100 mm^3. The growth rate of a tumor line was monitored by comparing the doubling times of a series of passages.

Results

Using light microscopy, 4 out of 48 consecutive samples of ovarian cancer patients contained fibrous or fatty tissue without vital tumor cells and were excluded from the study. Positive takes from samples of the primary tumor or a metastatic lesion (omentum or peritoneum) were 2 out of 16 (13%) and 12 out of 28 (43%), respectively. The overall take rate was 32% (14/44). Eventually 5/44 samples (11%) could be grown as tumor lines with passages beyond 3. In the stepwise logistic regression analysis the transplantability appeared not only to be dependent on the origin of the tumor sample ($p < 0.05$), but also on the number of mitotic figures ($p < 0.01$). The take rate was further dependent on both the number of mitotic figures and the degree of lymphocytic infiltration ($p < 0.02$). These conclusions formed the basis to draw up Table 3. The highest chance of success in growing human ovarian cancer xenografts was provided by the use of patients' samples derived from metastatic lesions containing more than 5 mitotic figures in 5 high-power fields and without lymphocytic infiltration. The other variables were not of influence on the transplantability of the tumor sample.

Table 3. Take rate dependency on the origin of the tumor sample, the number of mitotic figures and the degree of lymphocytic infiltration

Origin	Mitotic figures	Lymphocytic infiltration	Take rate[a] (%)
all samples	≤5	all samples	2/19 (11)
primary tumor	>5	moderate/severe	0/4 (0)
metastasis	>5	moderate/severe	3/9 (33)
primary tumor	>5	none/minimal	2/4 (50)
metastasis	>5	none/minimal	7/8 (88)

[a] numerator: number of positive takes; denominator: number of attempts.

Table 4. Histochemistry and CEA in human ovarian cancer lines

Tumor line	PAS	PAS/diastase	AB	CEA
Ov. Pe	++	++	++	+
MRI-H-207	+	−	−	−
Ov. Me	+	−	+/−	−
Ov. He	++	++	++	+
Ov. GI	+/−	−	+/−	−
Ov. Ri(C)	+/−	−	+/−	−
FCo	+	+/−	+	−
FKo	+	+/−	+/−	−
FMa	+/−	−	+/−	−
A 2780S	+/−	−	−	−

All tumor lines currently in use for chemotherapy studies were derived from various histologic subtypes and invariably retained this pattern throughout serial passages; also tumor lines Ov. Pe and Ov. He continued producing CEA (Tables 2 and 4).
In the electrophoretic separation of LDH fractions from human tumor xenografts, human LDH isoenzymes could invariably be distinguished, because human LDH-IV and -V moved further to the cathode than any of the murine fractions. Thus far, those tumor lines that could be processed easily for visible metaphases contained human chromosomes.
Comparison of the growth rate of various passages in the same tumor line did not reveal gross changes. It is important to note that the T_D of the lines differed with periods from 3.5 to 18 days.

Discussion

In our study using human ovarian cancer tissue transplanted subcutaneously in nude mice, the initial take rate was 32%, and 11% of samples could be grown as tumor lines. Thus far, the tumor lines have retained the histologic pattern and biologic characteristics of origin to a great extent.

The establishment of human ovarian cancer lines is very difficult. Kullander et al (5) transplanted 15 ovarian cancer samples and observed an initial take rate in 10 (67%) instances. Only 3 transplants (20%) were accepted in serial transfer. Friedlander et al (3) mentioned a take in 13 of 23 (57%) ovarian cancer samples of which 6 (26%) could be successfully transferred into subsequent passages. For ovarian cancer several factors appear to be related with the transplantability. We demonstrated that samples derived from a metastatic lesion with a high number of mitotic figures and absence of mononuclear infiltration were best transplantable regardless of their histologic subtype or degree of differentiation. In addition, host factors may also play a role in the rejection of heterologous malignant tissue. In nude mice pre-treated with cyclophosphamide we found an enhanced take rate in some of our lines, which may have been caused by reduction of active natural killer cells responsible for inhibition of tumor growth (6).

Tumor lines should be monitored continuously for retention of original characteristics. Parameters may change, such as the occurrence of a mouse sarcoma invading the human tissue (7) or the acceleration of growth rate in subsequent passages (8) having a negative impact on chemotherapy results.

At present we have at our disposition a series of human ovarian cancer lines differing in histologic subtype and growth rate. These differences may also be expected in a group of ovarian cancer patients. The selective transplantability of ovarian cancer samples may indicate that our panel does not comprise all (biologic) aspects of this tumor type. Comparative studies of drug effects in the panel and results of single agents in patients should give insight in the reliability of the model for the prediction of the activity of new agents in the clinical situation.

References

1. Houghton JA and Taylor DM: Maintenance of biological and biochemical characteristics of human colorectal tumours during serial passage in immune-deprived mice. Br J Cancer 1978 (37): 199-212
2. Kovnat A, Buick RN, Connolly JG, Jewett MA, Keresteci AG and Tannock IF: Comparison of growth of human bladder cancer in tissue culture or as a xenograft with clinical and pathological characteristics. Cancer Res 1984 (44): 2530-2533
3. Friedlander ML, Russel P, Taylor IW and Tattersall MHN: Ovarian tumour xenografts in the study of the biology of human epithelial ovarian cancer. Br J Cancer 1985 (51): 319-333
4. Boven E: Human ovarian xenografts in nude mice. Application of a new model in the selection of novel agents for ovarian cancer. Amsterdam: Rodopi 1986
5. Kullander S, Rausing A and Trope C: Human ovarian tumors heterotransplanted to "nude" mice. Acta Obstet Gynecol Scand 1978 (57): 149-159
6. Nauta MM, Boven E, Schlueper HMM, Erkelens CAM and Pinedo HM: Enhanced transplantability of human ovarian cancer lines in cyclophosphamide-pretreated nude mice. Br J Cancer 1986 (54): 331-335
7. Beattie GM, Knowles AF, Jensen FC, Baird SM and Kaplan NO: Induction of sarcomas in athymic mice. Proc Natl Acad Sci USA 1982 (79): 3033-3036
8. Fodstad Ø, Rofstad EK, Tveit KM and Pihl A: Spontaneous alteration in growth rate of two human melanoma xenografts. Concurrent change in chemosensitivity. Eur J Cancer Clin Oncol 1983 (19): 1175-1178

Representativity of Xenografts for Clinical Cancer. Tumor and Host Characteristics as Variables of Tumor Take Rate

Ø. FODSTAD

Department of Tumor Biology, Institute for Cancer Research, The Norwegian Radium Hospital, Montebello, 0310 Oslo 3, Norway

Abstract

Success in establishing growth of human tumors in nude mice depends on the characteristics of the tumor as well as those of the host. Careful selection and handling of the tumor tissue is important. Some tumor types are particularly difficult to grow in nude mice, but even for cancers that can be established more easily the representativity of xenograft panels for their clinical counterparts should be carefully considered. For melanomas a parameter of malignancy of the parent tumors was found to be reflected in the growth rates of the xenografts. Tumor heterogeneity represents a complex problem, and conclusions based on single parameters of heterogeneity should be drawn with great caution. By and large, xenografted tumors show a remarkable stability upon long-term passaging in the mice. The degree of immunodeficiency of the animals may affect tumor take rates. Age, health and immunoconstitution are factors of importance. Subcutaneous inoculation of tumor tissue is practical for most purposes, but "orthotopic" growth may be more clinically relevant for some tumors. New metastasis models have become available that permit biology and chemosensitivity studies to be performed under conditions closely similar to the situation in patients.

Introduction

Since the first successful transplantation of a human tumor into nude mice (1), a long series of experimental studies have provided evidence that xenografted tumors retain the most important biological features of the tumor of origin (2-4). Moreover, the morphological and biochemical characteristics seem by and large to remain stable upon serial transplantation in the new host (4, 5). Within each histological tumor type the xenografted lines have been found to display a range of different growth rates and chemosensitivities (6-9), reflecting the situation in the clinic. For many purposes, therefore, human tumor xenografts seem to provide a nearly optimal experimental model. In addition, they represent a unique tissue bank permitting an indefinite number of studies to be performed on the same untreated human tumor tissue.

In spite of these important and advantageous features, questions can still be raised as to what extent panels of xenografted tumors represent the complete range of clinical cancers of the respective histological types. It is not clear whether a selection of tumor cell subpopulations may take place during adaptation to the new environment. Tumor take rate is known to be dependent on tumor type (3, 10, 11) as well as on the immunological status of the host (11-14). These issues are addressed in the present paper, and their implications for the practical use of xenografts are discussed in relation to some relevant experimental data.

Tumor-Related Factors Influencing Take Rate and Growth of Xenografts

Tumor Sampling

Careful selection and handling of the tissues to be transplanted is clearly of utmost importance for obtaining take of xenografted tumors. Necrotic tissue must be discarded, whereas tissue from the periphery of the tumor may contain the highest proportion of proliferating cells. The selection of tissue containing viable tumor cells is particularly difficult in cancers with a high fraction of fibrotic tissue surrounding the malignant cells, e.g. in breast carcinomas. In colorectal cancer precautions must be taken to prevent bacterial contamination. The tumor tissue should be kept cold in sterile saline or medium until implantation, which should take place shortly after biopsy. If kept refrigerated, some delay in tumor transplantation is not critical. In general there seems to be a tendency towards obtaining higher take rates after subcutaneous implantation of small tumor pieces (approximately $2 \times 3 \times 3$ mm) than after injection of a suspension of single cells prepared from the same tumor.

Tumor Type

During recent years attempts have been made to grow cancers of almost any type as xenografts in nude mice. A significant difference in take rate has been observed between tumors belonging to different histological groups. The highest take rates are reported for malignant melanoma, colon and lung carcinomas (4, 7, 9-11, 15, 16). Intermediate success rates are obtained with ovarian, kidney and head and neck tumors (11, 16-20), whereas breast, prostate and testicular cancers are difficult to establish as permanent tumor lines (9, 11, 16, 21, 22). Finally, only very few cases of leukemia and lymphoma xenografts have been described (23-26). Preconditioning of the host has been shown to increase the take of some breast cancers, lymphomas and leukemias (25-27), but such treatment cannot significantly increase the overall take rate of these tumors. A medium-sized, well vascularized metastasis is usually a better tissue source than a large primary with a low growth fraction. Within each tumor type the take rates seem to correlate with functional and histological signs of tumor malignancy (11).
The mechanism behind the differences in take rate are poorly understood. For several tumor types it will therefore remain difficult to establish panels of xenografts that are representative of their clinical counterparts. We have asked whether insufficient representativity could constitute a problem even for a tumor type with a high growth potential in nude mice, like melanoma. In Fig. 1 growth curves for 6 xenografted human melanoma lines are shown. The tumor volume doubling times (30-60 mm^3) varied from 2.0 to 10.4 days, a variation in growth rates that could possibly reflect the range of growth differences known to exist for clinical melanomas. In order to elucidate the relationship between tumor growth rate in patients and in nude mice we have listed some parameters of tumor progression in patients, together with the volume doubling times of the respective xenografts in nude mice (Table 1). In the patients the time from diagnosis to first relapse of the disease, or to the appearance of distant metastasis, varied over a wide range. No relationship between these parameters and the growth rate of the xenografts could be found. However, when the time from first relapse to death of the patients was compared with tumor growth rate in mice, a correlation was seen. Since our xenografts originate from tumor biopsies of patients in relapse, it seems reasonable to conclude that the growth rate of melanomas in nude mice reflects the inherent growth capacity of the parent tumor. It is reported, however, that only 55-86% of melanomas can be grown as xenografts (9-11). Furthermore, we have observed that some melanoma xenografts that can be serially passaged, tend to grow extremely slowly and may never reach a diameter of more than 4-5 mm. Therefore, even for malignant melanoma it may be difficult to establish test panels of xenografts that a

Table 1. Relationship between parameters of malignancy in melanoma patients and the growth rate of the corresponding xenografts in nude mice

Designation	Time (months) from appearance of			Volume doubling time of xenograft[a] (days)	Melanin content[b]
	Primary to first relapse	Primary to distant metastasis	First relapse to death		
LO	3	4	5	2.0	–
SES	32	32	1	3.0	–
EE	6	8	5.5	3.6	–
KF	60	66	10	4.0	–
GRE	18	18-20	13	4.8	+/–
EF	18	22	13	6.3	+
RV	13	21	17	6.8-8.0	–
KA	120	124	16	8.0 (4.2)[c]	++(+)
NSR	144	144	14	8.0	–
MF	3-4	10-20	15	8.9	(+)
GE	41	41	32	9.0	–

[a] Time required to double mean tumor volume (50-100 mm^3).
[b] The melanin content was the same in the biopsy and in early xenograft passages.
[c] Tumor lines established from 2 different metastases showed different growth rates in mice.

priori can be regarded as truly representative of the clinical tumors. The implications of this conclusion depend on the purpose for which the xenografts are being used, but this possible limitation of the model should be taken into consideration when experimental results are evaluated.

It is our experience that the degree of melanin production in the original tumor is retained in the xenograft. The results in Table 1 show that when different tumors are compared, no correlation can be seen between the melanin content and tumor growth rate. However, when sublines of a particular tumor, growing at different rates, are studied, we have observed that such a correlation prevails.

Tumor Heterogeneity

In the same patient cancers may exhibit intratumoral as well as intertumoral heterogeneity. Tumor pieces taken from different parts of a biopsy have in our experience never given rise to xenografts with

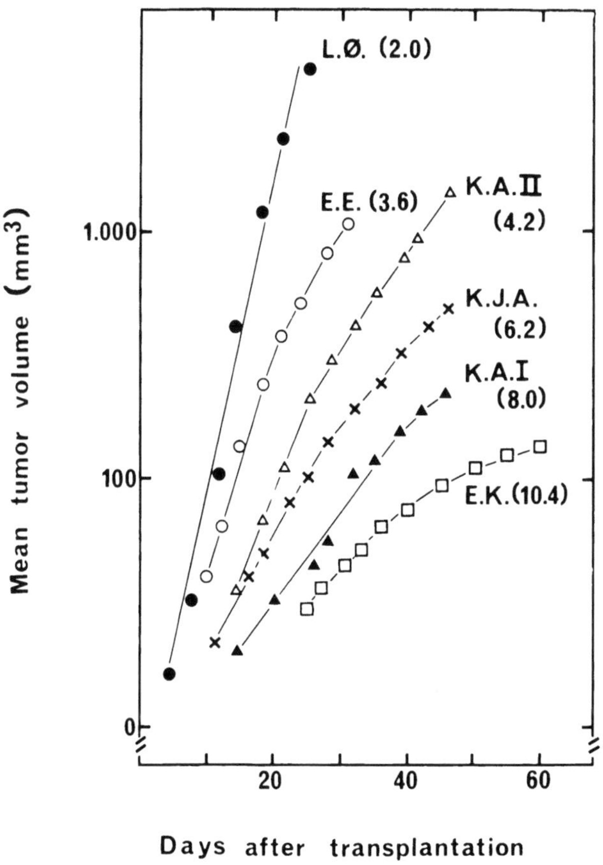

Fig. 1. Growth curves for 6 xenografted human melanoma lines in nude mice. For each tumor line the size of 6-8 s.c. xenografts in individual mice was measured twice weekly, and the mean tumor volumes were calculated according to the formula $\frac{1}{6}\pi$ (mean diameter)^3. The tumor volume doubling times (30-60 mm^3) are given in parenthesis

significant differences in growth characteristics upon serial passaging. In contrast, xenografts established from different metastases of the same patient may show clearly different growth rates, as illustrated by the tumors K.A. I and K.A. II in Fig. 1.

Heterogeneity within a xenografted tumor would be expected to result in a tumor line that, due to selection pressure in the host, would change phenotype rather easily. However, human tumor xenografts show, with few exceptions, a remarkable stability in characteristics during long term passaging (4, 5). Thus, some of our lines have remained unchanged during up to 10 years of growth in nude mice. In 2 melanomas, however, the growth rates changed spontaneously, followed by concurrent changes in chemosensitivity (28). The sensitivity profile for one of the lines, THX, recently changed for a second time, possibly reflecting an inherent high lability of this tumor. DNA-histograms obtained on tumor lines are being used to detect intratumoral heterogeneity. It is noteworthy that the histogram for the THX line, indicating 2 separate subpopulations of tumor cells, has remained the same in spite of the drastic biological changes that have taken place in this tumor.

It should be remembered that alterations in some major tumor characteristics may not necessarily be due to a selection of preexisting subclones, or reflect profound changes in the genome, but rather be caused by epigenetic factors induced by differences in microenvironment, thus representing what has been described as "dynamic heterogeneity" (29). Indications that this sometimes may be the case is provided by the following examples. One of our diploid melanoma lines changed spontaneously to become near tetraploid. However, during 3 months of further passaging the tumor reverted to its original karyotype. This has remained stable for 4 years. Similarly, major differences in melanin content that developed spontaneously in 2 xenografts of the same line growing in different mice, were retained in a few new passages. But a pronounced variation in melanin production was later observed in both "sublines". Moreover, in xenografted melanoma lines we have observed inconsistent alterations in hormonal immunoreactivity with increasing passage generations (30).

Site of Implantation

The subcutaneous (s.c.) space has for obvious reasons been preferred for growing human tumors in nude mice. It is technically easy to prepare a "bed" for the tumor inocula, and the xenografts can readily be observed through the thin, hairless skin. Some investigators have found that the anterior lateral thoracic wall is a better transplantation site than the posterior part of the trunk (31). We have not observed such site-dependent growth differences. We feel that for practical reasons it is favourable to have the tumor in the flank of the animal, where it does not cause unnecessary problems for the mouse, and it makes it easy for the investigator to measure tumor size. A possible disadvantage of growing tumors s.c. is that the blood supply, provided through the vasculature in the fibrous capsule (4), is relatively poor.

Inoculation of tumor cells intraperitoneally (32-34), intracerebrally (23, 35), intrasplenically (35), and intravenously (14, 32, 37) has also been performed. In some cases these procedures may provide certain advantages with respect to take rate, but the available data do not permit strong general conclusions to be drawn. Use of the mammary fat pad has been reported to facilitate growth of some breast cancers, possibly due to the hormonal milieu provided (22). Recently, "orthotopic" inoculation of a renal cell carcinoma into the kidney (38) and of lung cancers intrabronchially (39), has been used in order to obtain models that more closely mimic the situation in patients. These procedures provide interesting possibilities, but tumor type specific growth advantages have not yet been demonstrated.

Host Factors Influencing Take Rate of Xenografts

The lack of thymus renders nu/nu mice devoid of functional lymphocytes, but in some cases T-cell-like activity has been detected, possibly after some type of stimulation of immature cells (40, 41). Moreover, nude mice do have normal or high levels of other immune mechanisms believed to be engaged in host defense against tumors (42), mechanisms involving B cells, NK cells and macrophages. Factors that further increase the immunodeficiency of athymic mice, therefore, may help increase the take rate of xenografted human tumors.

Age

Newborn and young, 2-4 week-old, nude mice have been shown to be immunologically immature (14), and tumor take rate as well as the extent of metastasis formation may be increased in these animals relative to the situation in adult mice (13, 43). Since low NK-cell activity has been associated with increased tumor take (13, 43, 44), and the NK function is low in young nude mice (13, 14), the increased tumor growth in 2-4 week-old nudes has been attributed to their reduced level of NK activity (13, 43). It should be remembered that, in addition to being NK deficient, these animals have impaired function of other immune mechanisms as well (14).

Health

It is well known that the T-cell deficiency makes nude mice prone to infections caused by viruses and other microorganisms that do not constitute any threat to conventional rodents. Of these infections hepatitis is particularly deleterious. However, even much less dramatic infections can affect the host immune system in a way that may influence the growth of transplanted tumor tissue. Low virulent infections may possibly stimulate part of the immune system to become more efficient in preventing take of xenografts.

The growth rate of transplanted tumors is decreased in animals that, due to infections, are in a poor general condition, associated with a negative nutritional balance. Furthermore, infections may shorten the life span of the mice to such an extent that tumors with low growth capacity and/or long lag times, may not have time to get established as progressively growing tumors. In conclusion, a healthy animal stock is of major importance for success in transplanting human tumors in nude mice.

Conditioning of the Mice

Irradiation and anti-thymocyte serum (25, 32) as well as anti-asialo GM 1 antibody (45) are examples of pretreatment that may increase the responsiveness of nude mice to tumor transplants. None of these procedures, however, act specifically on one particular population of immune cells, and the treatment may have effects on the experimentation to be performed.

Immunoconstitution

The nu/nu homozygosity has been introduced into mice of different outbred and inbred genetic backgrounds. Also double immunodeficient animals have been bred that are defective in either B-cell or NK-cell activity, in addition to their lack of active T cells (12, 14, 46). Improved take rate in B-cell deficient NIH II mice was initially reported (12), but this was not confirmed in a later study (14). Moreover, in contrast to what was expected (43, 44), beige-nude mice with low levels of NK cell activity (46) were not found to be superior to regular nudes with regard to tumor take (14). Recently, triple immunodeficient NIH III mice have been developed that have low activity levels of T, B and NK cells. In these mice the take rate of some slowly growing tumors has been found to be increased relative to that in regular nudes (47).

Metastasis of Human Tumors in Nude Mice

Metastasis formation is a rather rare event in nude mice carrying s.c. xenografts which usually grow locally in a semi-benign fashion (32). Since metastatis constitutes the major therapeutic problem in the clinic, the validity of therapy experiments in such models could be questioned. However, the s.c. xenograft model has proven its usefulness, in spite of being suboptimal with regard to some aspects. Nevertheless, it is of great interest that several metastatic tumor models have recently been developed in nude mice (14, 32, 34, 37, 38, 48-50). These models allow different aspects of the metastatic process to be studied, and they provide possibilities for testing effects of therapy on the process of tumor metastasis as well as on the growth of metastases established in clinically relevant tissues.
The LOX melanoma (14, 37, 51) has shown lung tissue specificity after injection of cells i.v. (14, 37) and lung metastases are formed when sublines are growing subcutaneously (52) or as ascites (34, 52). Other tumors are reported to give rise to metastases in various tissues, partly dependent on the site of inoculation of the cells (32, 36, 38, 50). In our laboratory we have recently established several tumor lines, including sarcomas and melanomas, that after i.v. injection of single cells develop metastasis in organs typical for the metastatic pattern seen for these tumor types in patients. One example is the melanoma line FEMX M I which originates from a lymph node in the patient. Upon i.v. injection into nude mice, cells from this line exclusively give lymph node or s.c. metastasis.
The availability of relevant metastatic models opens new perspectives for the study of human tumor biology, and they provide valuable tools in the attempts to obtain improved chemotherapy of cancer.

Acknowledgements

The author thanks Dr. Alexander Pihl for invaluable discussions, Mrs. Unni Ronning and Ms. Marianne Isaksen for expert technical assistance, and Mrs. Eva Skaar and Ms. Ingvild Eggen for typing the manuscript.

References

1. Rygaard J and Povlsen CO: Heterotransplantation of a human malignant tumor to "nude" mice. Acta Path Microbiol Scand 1969 (77): 758-760
2. Fodstad Ø, Olsnes S and Pihl A: Inhibitory effect of abrin and ricin on the growth of transplantable murine tumors and of abrin on human cancers in nude mice. Cancer Res 1977 (37): 4559-4567
3. Giovanella BC, Stehlin JS, Fogh J and Sharkey FE: Serial transplantation of human malignant tumors in nude mice and their use in experimental chemotherapy. In: Houchens D and Ovejera T (eds) Proceedings of the symposium on the use of athymic (nude) mice in cancer research. New York: Gustav Fischer 1978: 163-179
4. Povlsen CO, Rygaard J and Fogh J: Long-term growth of human tumors in nude mice: Evaluation of stability. In: Fogh J and Giovanella BC (eds) The nude mouse in experimental and clinical research. New York: Academic Press 1982 (2): 79-93
5. Sordat B, Fritsche R, Mach P, Carrel S, Ozzello L and Cerotini JC: Morphological and functional evaluation of human solid tumors serially transplanted in nude mice. In: Rygaard J and Povlsen CO (eds) Proceedings of the first international workshop on nude mice. Stuttgart: Gustav Fischer 1974: 269-277
6. Fodstad Ø, Aass N and Pihl A: Response to chemotherapy of human malignant melanoma xenografts in athymic nude mice. Int J Cancer 1980 (25): 453-458
7. Shorthouse AJ, Peckham MJ, Smyth JF and Steel GG: The therapeutic response of bronchial carcinoma xenografts: A direct patient-xenograft comparison. Br J Cancer 1980 (41, suppl. IV): 142-145
8. Giovanella BC, Stehlin JS Jr, Shephard RC and Williams LJ: Correlation between response to chemotherapy of human tumors in patients and in nude mice. Cancer 1983 (52): 1146-1152
9. Steel GG, Courtenay VD and Peckham MJ: The response to chemotherapy of a variety of human tumor xenografts. Br J Cancer 1983 (47): 1-13
10. Fiebig HH, Schuchhardt C, Henss H, Fiedler L and Löhr GW: Comparison of tumor response in nude mice and in the patients. Behring Inst Mitt 1984 (74): 343-352
11. Sharkey FE and Fogh J: Considerations in the use of nude mice for cancer research. Cancer Metastasis Rev 1984 (3): 341-360
12. Azar HA, Hansen CT and Costa JN: N:NIH (S) II-nu/nu mice with combined immunodeficiency. A new model for human tumor transplantations. J Natl Cancer Inst 1980 (65): 421-430
13. Hanna N and Fidler IJ: Expression of metastatic potential of allogeneic and xenogeneic neoplasms in young nude mice. Cancer Res 1981 (41): 438-444
14. Fodstad Ø, Hansen CT, Cannon GB, Statham CN, Lichtenstein GR and Boyd MR: Lack of correlation between natural killer activity and tumor growth control in nude mice with different immune defects. Cancer Res 1984 (44): 4403-4408
15. Mattern J, Jäger S, Sonka J, Wayss K and Volm M: Growth of human bronchial carcinomas in nude mice. Br J Cancer 1985 (51): 195-200
16. Fogh J, Fogh JM and Orfeo T: One hundred and twenty-seven cultured human tumor cell lines producing tumors in nude mice. J Natl Cancer Inst 1977 (59): 221-226

17. Kullander S, Rausing A and Tropé C: Human ovarian tumours heterotransplanted to 'nude' mice. Acta Obstet Gynecol Scand 1978 (57): 149-159
18. Boven E: Human ovarian cancer xenografts in nude mice. Application of a new model in the selection of novel agents for ovarian cancer. Amsterdam: Rhodopi 1986
19. Nauta MM, Boven E, Schlüper HMM, Erkelens CAM and Pinedo HM: Enhanced transplantability of human ovarian cancer lines in cyclophosphamide-pretreated nude mice. Br J Cancer 1986 (54): 331-335
20. Braakhuis BJM, Sneeuwloper G and Snow GB: The potential of the nude mouse xenograft model for the study of head and neck cancer. Arch Othorhinolaryngol 1984 (239): 69-79
21. Rae-Venter B and Reid LM: Growth of human breast carcinomas in nude mice and subsequent establishment in tissue culture. Cancer Res 1980 (40): 95-100
22. White AC, Levy JA and McGrath CM: Site-selective growth of a hormone-responsive human breast carcinoma in athymic mice. Cancer Res 1982 (42): 906-912
23. Epstein AL, Herman MM, Kim H, Dorfmann RF and Kaplan HS: Biology of the human malignant lymphomas. III. Intracranial heterotransplantation in the nude, athymic mouse. Cancer 1976 (37): 2159-2176
24. Adelmann DC, Miller RA and Kaplan HS: Humoral immune response to human lymphoma cell heterotransplants in the central nervous system of athymic, nude mice. Int J Cancer 1980 (25): 467-473
25. Watanabe S, Shimosato Y, Kuroki M, Sato Y and Nakajima T: Transplantability of human lymphoid cell line, lymphoma and leukemia in splenectomized and/or irradiated nude mice. Cancer Res 1980 (40): 2588-2595
26. Ziegler HW (Löms), Frizzera G and Bach FH: Successful transplantation of a human leukemia cell line into nude mice: Conditions optimizing graft acceptance. J Natl Cancer Inst 1982 (68): 15-18
27. Leung CKH and Shin RPC: Required presence of both estrogen and pituitary factors for the growth of human breast cancer cells in athymic nude mice. Cancer Res 1981 (41): 546-551
28. Fodstad Ø, Rofstad EK, Tveit KM and Pihl A: Spontaneous alternation in growth rates of two human melanoma xenografts. Concurrent changes in chemosensitivity. Eur J Cancer 1983 (19): 1175-1178
29. Harris JF, Chambers AF, Hill RP and Ling V: Metastatic variants are generated spontaneously at a high rate in mouse KHT tumor. Proc Natl Acad Sci USA 1982 (79): 5547-5551
30. Gould VE, Fodstad Ø, Memoli VA, Warren WH, Dardi LE and Johannessen JV: Neuroendocrine cells and associated neoplasms of the skin. In: Falkmer S, Håkanson R and Sundler F (eds): Evolution and tumor pathology of the neuroendocrine system. Amsterdam: Elsevier 1984: 545-580
31. Kyriazis AA and Kyriazis AP: Preferential sites of growth of human tumors in nude mice following subcutaneous transplantation. Cancer Res 1980 (40): 4509-4511
32. Sordat BCM, Ueyama Y and Fogh J: Metastasis of tumor xenografts in the nude mouse. In: Fogh J and Giovanella BC (eds) The nude mouse in experimental and clinical research 1982 (2): 95-147
33. Hamilton TC, Young RC, Louie KG, Behrens BC, McKoy WM, Grotzinger KR and Ozols RF: Characterization of a xenograft model of human ovarian carcinoma which produces ascites and intraabdominal carcinomatosis in mice. Cancer Res 1984 (44): 5286-5290
34. Shoemaker R, Wolpert-DeFilippes M, Mayo J, Abbott B, Plowman J, Venditti J, Boyd M, Fodstad Ø, Ovejera A, Donovan P, Griswold D and Dykes D: Experimental chemotherapy studies of a human melanoma in nude mice using a survival endpoint. Proc Amer Assoc Cancer Res 1985 (26): 330
35. Schold SC, Rawlings CE, Bigner SH and Bigner DD: Intracerebral growth of a human glioma tumor line in athymic mice and treatment with procarbazine, 1,3-Bis (2-chloroethyl)-1-nitrosourea, aziridinyl-benzoquinone, and cis-platinum. Neurosurgery 1983 (12): 672-677
36. Giavazzi R, Campbell DE, Jessup JM, Cleary K and Fidler IJ: Metastatic behavior of tumor cells isolated from primary and metastatic human colorectal carcinomas implanted into different sites in nude mice. Cancer Res 1986 (46): 1928-1933
37. Fodstad Ø, Aamdal S, Tveit KM and Pihl A: Lung colony formation in adult nude mice upon intravenous injection of cells from a human melanoma xenograft. Proc Amer Assoc Cancer Res 1982 (23): 223
38. Naito S, von Eschenbach AC, Giavazzi R and Fidler IJ: Growth and metastasis of tumor cells isolated from a human renal cell carcinoma implanted into different organs of nude mice. Cancer Res 1986 (46): 4109-4115
39. McLemore TL, Blacker PC, Gregg M, Jessee SE, Alley MC, Abbott BJ, Shoemaker RH, Litterst CC, Hubbard WC, Brennan RN, Fine DL, Eggleston JC, Mayo JG and Boyd MR: Intrabronchial implantation: A method for the orthotopic propagation of human lung tumors in athymic nude mice. Cancer Res 1987 (47): 5132-5140
40. MacDonald HR: Phenotypic and functional characteristics of "T"-like cells in nuce mice. Exp Cell Biol 1982 (50): 305
41. Wagner H, Hardt C, Heeg K, Röllinghof M and Pfizenmaier K: T-cell derived helper factor allows in vivo induction of cytotoxic T cells in nu/nu mice. Nature (Lond) 1980 (284): 278-280
42. Findred B: Nude mice in immunology. Prog Allergy 1979 (26): 137-238
43. Hanna N: Role of natural killer cells in control of cancer metastasis. Cancer Metastasis Rev 1982 (1): 45-64
44. Herberman RB and Holden HT: Natural cell-mediated immunity. Adv Cancer Res 1978 (27): 305-377
45. Kasai M, Yoneda T, Habu S, Maruyama Y, Okumura K and Tokunaga T: In vivo effect of anti-asialo GM1 antibody on natural killer activity. Nature (Lond) 1981 (291): 334-335
46. Fodstad Ø, Hansen CT, Cannon GB and Boyd MR: Immune characteristics of the beige-nude mouse. Scand J Immunol 1984 (20): 267-272
47. Fodstad Ø, Hansen CT, Cannon GB, Aamdal S and Boyd MR: Triple-immunodeficient nude mice. An improved model for growing human tumor xenografts (Manuscript in preparation)
48. Kerbel RS, Man MS, Dexter D: A model for human cancer metastasis of a tumor pigmented melanoma and derived variant sublines in nude mice. J Natl Cancer Inst 1984 (72): 93-108
49. Kuroki M: A new human osteosarcoma strain (Os-Beppu) with high rate of pulmonary metastasis in nude mice. Jpn J Clin Oncol 1984 (14): 49-56

50. Kozlowski JM, Fidler IJ, Campbell D, Xu XL, Kaighn ME and Hart IR: Metastatic behavior of human tumor cell lines grown in the nude mouse. Cancer Res 1984 (44): 3522-3529
51. Fodstad Ø, Aamdal S, Pihl A and Boyd MR: Activity of mitozolomide (NSC 353451), a new imidazotetrazine, against xenografts from human melanomas, sarcomas, lung and colon carcinomas. Cancer Res 1985 (45): 1778-1786
52. Shoemaker RH, McLemore TL, Abbott BJ, Fine DL, Gorelik E, Mayo JG, Fodstad Ø and Boyd MR: Human tumor xenograft models for use with an in vitro based, disease-oriented antitumor drug screening program. In: Winograd B, Peckham MJ and Pinedo HM (eds) Human tumour xenografts in anticancer drug development. ESO Monographs. Heidelberg: Springer 1988: 115-122

II. Prediction of Tumour Response for the Individual Patient by the Human Tumour Xenograft

Comparison of Tumor Response in Nude Mice and in Patients*

H. H. FIEBIG

Department of Internal Medicine, Division of Hematology/Oncology, Hugstetter Straße 55, 7800 Freiburg i. Br., Federal Republic of Germany

Abstract

Four hundred and six human tumors of different origin were transplanted subcutaneously into nude mice. Histological evidence of viable tumor cells was obtained in 79%, rapid tumor growth after 3 months in 51%, and 49% of these tumors were transferred in serial passages. Most of the latter tumors were suitable for drug testing. In order to investigate whether tumor response in nude mice is consistent with patient response, 80 comparisons were performed comprising 55 tumors. Comparisons included carcinomas of the large bowel (27), lung (13), stomach (9), melanomas (8), breast (4) and 19 other tumors. Combination chemotherapy was more successful than single agent therapy (37 vs 14% remission rate). In 21 patients tumors went into remission, consistent with 19 cases in the nude mouse system. In 59 cases the patient did not respond to therapy, corresponding to 57 cases in the nude mouse system. Xenografts enabled correct predictions for resistance in 96% and for tumor response in 90% of cases. Despite great efforts to obtain a large number of comparisons, only 32 test results were available before the patients needed treatment. Therefore, our study has shown that the xenograft system will not have practical significance in determining individual patient treatment. Limitations are the duration of testing, the take rate of only 50% and the charges for nude mice. However, the highly correct prediction rates for tumor sensitivity and resistance validates human tumor xenografts as tumor models for testing new drugs and combinations. Furthermore, xenografts can be used as tumor source for experiments in cell culture.

* Dedicated to G. W. Löhr, Freiburg, on the occasion of his 65th birthday

Introduction

The response rates of the different human cancers to single agent and combination chemotherapy are well-known today. However, we cannot predict whether an individual patient will benefit from treatment. It might be possible to individualize drug regimens by testing a patient's tumor to a wide range of chemotherapeutic agents prior to treatment administration. For this purpose, several *in-vitro* "predictive assays" have been employed. They include long-term culture and evaluation of morphological changes (5), biochemical assays with inhibition of DNA or RNA synthesis (9), the *in-vitro* colony formation (6, 10), measurements of cell viability with dye exclusion assays (11) as well as reduction of tumor cells as measured by cytofluorometry (8). These *in-vitro* systems yielded highly correct prediction rates for tumor resistance. However, no system could obtain clinical significance in selecting drugs for individual patient treatment.

Over the past 10 years, most of the solid human tumors have successfully been grown in athymic nude mice. For chemotherapy experiments this *in-vivo* system presents advantages in comparison to *in-vitro* drug exposure. Drugs which must be activated or detoxified can be tested, and also drug combinations can be evaluated. During the past 6 years we have transplanted more than 400 different human tumors subcutaneously into nude mice. We will report on our take rates and the comparison of tumor response in nude mice and patients.

Methods

Animals

Athymic nude mice of the NMRI genetic background were used which were bred in our own nude mouse colony. The animals were kept in macrolon cages set in laminar flow rackets. They were maintained as described by Fortmeyer (4). In the first passage tumors from men were implanted into male nude mice, tumors from women into female mice. All therapeutic experiments with the exception of testicular cancers were carried out in female mice.

Tumors

Over the past 8 years, 406 resected human malignancies were implanted subcutaneously into nude mice. Tumor slices of approximately 5 × 5 × 0.5–1 mm diameter were implanted in the flanks of the animals, in the first passage usually 16 fragments into 4 nude mice. When the tumors reached diameters of 1–1.5 cm, they were subpassaged, and the remaining tumor material was studied histologically. In therapeutic experiments 2 tumor fragments were implanted into each animal.

Tumor Growth Measurements

Tumor growth was followed weekly by measuring 2 perpendicular diameters. The product of the 2 diameters was taken as a measure for tumor size. Relative tumor size values were calculated for each single tumor according to tumor size on day X divided by tumor size on day 0 at the time of randomisation, multiplied by 100. The median tumor size was taken for evaluation.

Experimental Design of Testing

Testing was performed in serial passages usually between passage 2 and 6 when the tumor growth became more regular. Mice were selected randomly for the untreated or the test groups after 3-6 weeks when individual tumors presented the following properties: product of the 2 tumor diameters ⩾ 10 mm², estimated depth of at least half of the smaller diameter. Tumors with a yellow color indicating a high amount of fibrous tissue were excluded. Using these criteria we never observed spontaneous regression or stationary growth behavior in the untreated control groups after 3-4 weeks.
At the time of randomisation the median product of the tumor diameters was 56 mm². Each test group consisted of 5-6 animals comprising between 6-10 evaluable tumors. At the time of randomisation several representative tumors were examined histologically. The experiment was completed when each single tumor had doubled its initial tumor size which was the case for the majority of the tumors after 4-10 weeks.

Table 1. Comparison of chemotherapy in nude mice and in the patient

Tumor type	No. of patients	Chemotherapy		
		Total	Combination	Single Drug
Colorectal	18	27	3	24
Lung, Small Cell	2	4	4	0
Non Small Cell	6	9	5	4
Stomach	6	9	6	3
Melanoma	7	8	7	1
Mammary	2	4	2	2
Sarcoma	3	4	4	0
Testicle	2	2	2	0
Bladder	1	2	1	1
Thymoma	1	2	2	0
Renal	1	2	1	1
Wilms Tumor	1	2	2	0
Miscellaneous	5	5	4	1
Total	55	80	43 (54%)	37 (46%)

Chemotherapy

Fifty-five patients were treated with established drug combinations (43 cases) or with single-agent chemotherapy (37 cases). After progression under the initial therapy a second-line therapy was administered in 25 cases. The tumor categories and the type of chemotherapy are shown in Table 1.
Patient follow up and evaluation of patient tumor response were performed as in clinical trials.
The treatment regimen in nude mice corresponded to clinical schedules, with the exception that treatment in mice was usually repeated after 2 weeks. Drugs, doses, schedules and route of administration are shown in Table 2. A dose around the LD10 after 14 days and around the LD20 after 28 days was considered to be the maximum tolerable dose in tumor-bearing immunodeficient nude mice. In immunocompetent non-tumor bearing mice 2 treatment courses with these dose schedules caused a lethality of less than 10% after 4 weeks. In combination chemotherapy drugs were administered at 15-minute intervals by different routes to avoid direct interactions of the drugs. In 2-drug combinations only 70-80% of the dose that would be used in single drug therapy could be given, and, accordingly, only 50-60% in 3-drug combinations.

Evaluation Parameters for Tumor Response

The tumors in nude mice were evaluated after maximum tumor regression, or after 3-4 weeks in non-regressing tumors. The effect of treatment was

Table 2. Maximum tolerable dose-schedules of anticancer drugs in tumor-bearing nude mice

Drug	Dose mg/kg/day	Schedule day	Route of administration	Day 14		Day 28	
				death/total	%	death/total	%
ACNU	20	1	i.p.	16/134	12	24/134	18
Adriamycin	8.0	1, 15	i.v.	22/267	8	58/260	22
CCNU	20	1	i.p.	33/197	17	50/217	23
Cisplatin	6.4	1, 15	s.c.	13/143	9	32/143	22
	8	1, 15	s.c.	19/ 98	19	30/ 98	31
Cyclophosphamide	200	1, 15	i.p.	18/180	10	41/180	23
DTIC	80	1-4, 15-18	i.p.	1/ 50	2	8/ 50	16
	300-350	1, 15	i.p.	5/ 53	9	11/ 53	21
Etoposid	24	1-3, 15-17	s.c.	17/123	14	23/123	19
Fluorouracil	40	1-4, 15-18	i.p.	28/232	12	45/232	19
	125	1, 8, 15	i.p.	4/ 32	13	7/ 32	22
Mitomycin-C	2.0	1, 15	i.v.	15/168	9	48/179	27
Vindesine	1.5	1, 8, 15	i.v.	15/299	5	33/294	11

ACNU = Nimustine
CCNU = 1-(2-chloroethyl)-3-cyclohexyl-1-nitrosourea
DTIC = Dacarbazine

classified in the xenograft system and in the patients as remission (product of 2 diameters < 50% of initial value), minimal regression (51-75%), no change (76-124%) and progression (≥ 125% of initial value). All patients had measurable lesions. Evaluation was usually performed after 2 treatment cycles or after maximal regression. The evaluation of tumor response in nude mice and in the patients was performed by different physicians.

Results

Take Rates and Growth Behavior

The take rates and the growth behavior of 406 different human malignancies are shown in Table 3. Histological evidence of viable tumor tissue was observed in 321 tumors (79%). Rapid tumor growth, defined as a tumor size of at least 60 mm^2 (a x b) after 90 days, was observed in 209 tumors (51%), and slow growth in 112 tumors (28%). Most of the rapidly growing tumors and in rare cases also initially slowly growing tumors were transferred to serial passages.

One hundred and ninety-seven malignant tumors (49%) were subpassaged at least 3 times. These high take rates were initially found in colorectal cancers only. After refining the technique, e.g. by selecting viable tumor tissue, similar take rates were observed in most of the other malignancies as well, with the exception of mammary cancers. One hundred and eighty regularly growing tumors of different tumor categories were selected as tumor models. Most of them were frozen in liquid nitrogen and are available for therapeutic and biological studies (1, 2).

Table 3. Transplantation of human tumors in nude mice. Take rate and growth behavior

Tumor type	Total number	Take[a]		Rapid[b] growth		Serial[c] passage	
Colorectal	94	73	78%	49	52%	53	56%
Stomach	46	32	70%	13	28%	14	30%
Lung -							
Non Small Cell	70	62	89%	47	67%	42	60%
Small Cell	13	11	85%	7	54%	8	62%
Sarcoma	33	30	91%	22	67%	18	55%
Melanoma	33	29	88%	19	58%	18	55%
Hypernephroma	30	20	67%	13	43%	10	33%
Miscellaneous	87	64	74%	39	45%	36	41%
Total	406	321	79%	209	51%	197	49%

[a] histologic evidence.
[b] tumor surface (a x b) ≥ 60 mm^2 after 90 days.
[c] at least 3 passages.

Comparison of Tumor Response

Fifty-five tumors allowed 80 comparisons between tumor response in nude mice and in patients. In colorectal cancers 24 comparisons were evaluated using single-agent chemotherapy with 5-fluorouracil or nitrosourea and 3 employing combination chemotherapy. Most of the other tumors were treated with combination chemotherapy (40 times);

Table 4. Comparison of tumor response to chemotherapy in nude mice and in the patient

Mouse/Patient	Total
Remission/Remission	19
No Remission/Remission	2
No Remission/No Remission	57
Remission/No remission	2

Table 5. Comparison of tumor response in nude mice and in the patient according to tumor category

Mouse/ Patient	Total	Colo- rectal	Stom- ach	Lung NSCLC	Lung SCLC	Mela- nomas	Breast	Sar- coma	Other[a]
R /R	19	2	4		4		2		7
NR /NR	57	23	5	8		7	2	4	8
NR /R	2	2							
R /NR	2			1		1			
Total	80	27	9	9	4	8	4	4	15

[a] 2 testicle, 2 bladder, 2 thymoma, 2 renal, 2 epidermoid of skin, 1 thyroid cancer, 2 Wilms tumor, 1 penile cancer, 1 ovarian cancer.

only 13 with single agent chemotherapy. Mammary cancers were initially treated with tamoxifen and upon progression with the combination of cyclophosphamide, methotrexate and fluorouracil. Stomach cancers were treated with the FAM combination (fluorouracil + adriamycin + mitomycin-C), and small-cell lung cancers with a combination of adriamycin, vincristine and cyclophosphamide, upon progression with cisplatin and VP-16.

The overall results are given in Table 4. A total of 21 patients went into remission, corresponding to 19 cases in the nude mouse. Fifty-nine patients did not respond to treatment, which was consistent with 57 cases of therapy resistance in the nude mouse system. Overall, xenografts gave correct predictions for resistance in 96% and for tumor response in 90%.

Combination chemotherapy was more successful than single-agent chemotherapy. Out of the 43 combinations given, 16 (37%) effected a remission in the patients, as compared to 5 out of 35 (14%) of single-agent chemotherapies. Single-agent therapy was successful in 4 patients with colorectal cancer and in 1 stomach cancer case. Tumor response according to tumor type is shown in Table 5. The growth curves of 2 xenograft tumors are shown in Figures 1 and 2. Details of comparative chemotherapy in individual tumors have been described recently for 55 comparisons (3).

Potential of the Nude Mouse System as Predicitive Assay

During a period of 5 years great efforts were made to obtain a large number of treatment comparisons. Approximately two thirds of the tumors were obtained during surgery of the primary tumor with no immediate need to treat the patient with chemotherapy. We tried to have the testing result in the nude mouse available before the patient needed treatment. Testing was performed between passage 2 and 6 when the tumor growth became more regular, so that the results became available at the earliest 4 months after surgery in rapidly growing tumors and 2 years after surgery in some cases of very slowly growing tumors.

In 28 cases the test result was available in the xenograft system before the patient needed chemotherapy, in 18 cases the tumor was treated simultaneously in the nude mouse and in the patient, and in 34 cases the patients were treated first (Table 6).

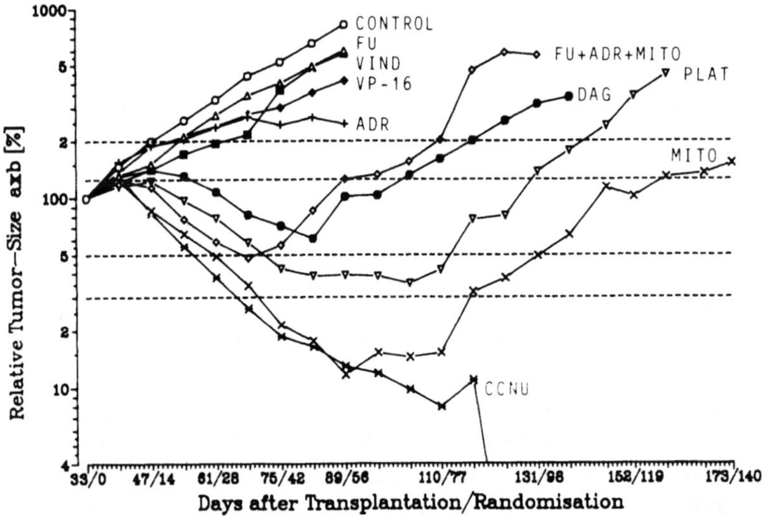

Fig. 1. Effect of single agent or combination chemotherapy on the growth of stomach cancer xenograft GXF 281. ADR = adriamycin; DAG = dianhydrogalactitol; FU = 5-fluorouracil; MITO = mitomycin-C; PLAT = cisplatin; VIND = vindesine; VP-16 = etoposid

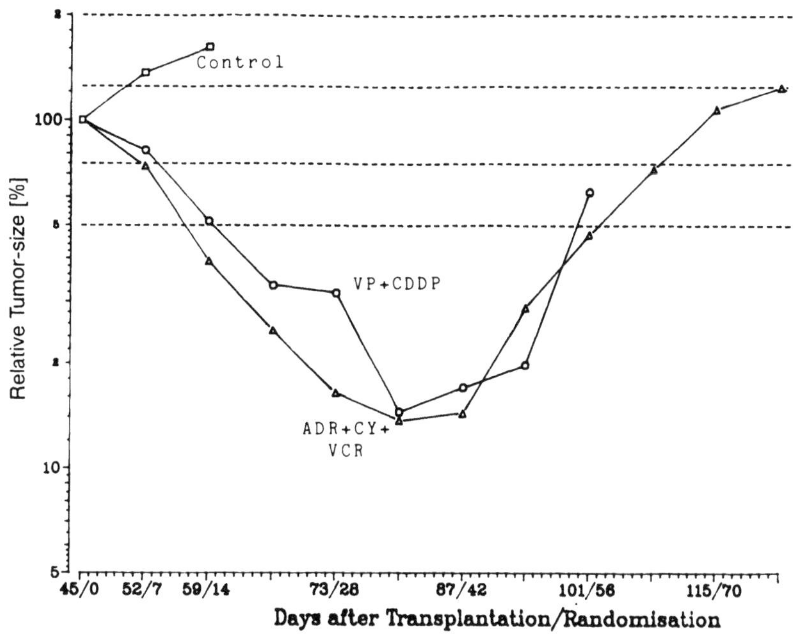

Fig. 2. Effect of single agent or combination chemotherapy on the growth of small-cell lung cancer xenograft LXF 177. □ = control; ○ = etoposid (VP) 24 mg/kg/day s.c. days 1-3, 15-17 and cisplatin (CDDP) 3.2 mg/kg/day s.c. days 1-2, 15-16; △ = adriamycin (ADR) 4 mg/kg/day i.v. days 1,15 and cyclophosphamide (CY) 150 mg/kg/day s.c. days 1,15 and vincristine (VCR) 0.5 mg/kg/day i.p. days 1,15

Table 6. Comparison of tumor response in nude mice and in the patient

	No.
Prospective[a]	28
Simultaneous	18
Retrospective[b]	34

[a] test result from nude mice available before patient needed chemotherapy.
[b] patient treated first.

Therefore, we can conclude that the xenograft system will not have practical significance in determining patient treatment.

Discussion

The main purpose of this study was to evaluate the significance of the xenograft system as a predictive assay for the individual treatment of patients and, furthermore, to validate the potential of the human tumor-nude mouse system for testing new drugs. Advantages include the highly correct prediction rates for resistance and sensitivity of a tumor. The testing is reproducible and also combinations and drugs which must be activated or detoxified can be studied. The highly correct prediction rates for sensitivity in comparison to *in-vitro* systems must be emphasized. In the stem cell assay and in biochemical assays only one half to two thirds of the drugs that were predicted to be effective were active in the clinic (6, 9). A better prediction for effective drugs was reported by Shorthouse et al. (7) with xenotransplantation into immunosuppressed mice.

However, our experience has shown that the nude mouse system will not have practical significance for predicting patient treatment. A major limitative factor is the duration of testing, requiring serial passage and thus a time period of at least 4 months in rapidly growing tumors, up to 2 years in slowly growing tumors, before test results can be obtained. In addition, nude mice are expensive; they need special conditions behind laminar flow barriers to avoid infections. Another limiting factor is the testing rate, since only half of the implanted tumors can be transferred to serial passages. Whereas the xenograft system cannot be used as a clinical routine method, the highly correct prediction rates for tumor sensitivity and resistance validate human tumor xenografts as human models for testing new drugs.

Acknowledgements

We wish to thank Prof. Dr. E.H. Farthmann and Prof. Dr. M. Schwaiger (Directors of the Chirurgische Universitätsklinik, Freiburg) for sending us human tumor tissue. We would like to thank our collaborators Miss C. Berg, Miss U. Dentler, Miss E. Hackenberg and Dipl.-Ing. Dr. K.H. Widmer for their competent assistance in this project.
This work was supported by grant PTB 8712 from the Bundesministerium für Forschung und Technologie.

References

1. Fiebig HH: Wachstum und Chemotherapie menschlicher Tumoren - vorwiegend Dickdarm-, Magen- und Bronchialkarzinome - in der thymusaplastischen Nacktmaus. Habilitationsschrift 1983, Universität Freiburg
2. Fiebig HH, Löhr GW: Wachstum menschlicher Malignome und Entwicklung von Tumormodellen in der thymusaplastischen Nacktmaus. Med Welt 1984 (35): 52-58, 81-86
3. Fiebig HH, Schuchhardt C, Henss H, Fiedler L, Löhr GW: Comparison of tumor response in nude mice and in the patients. Behring Inst Mitt 1984 (74): 343-352
4. Fortmeyer HP: Thymusaplastische Maus (nu/nu) - Thymusaplastische Ratte (rnu/rnu) - Haltung, Zucht, Versuchsmodelle. Schriftenreihe Versuchstierkunde 8. Berlin-Hamburg: Verlag P. Parey 1981
5. Limburg H, Krahe M: Züchtung von menschlichem Krebsgewebe in der Gewebekultur und seine Sensibilitätstestung gegen neuere Zytostatika. Dtsch med Wschr 1964 (89): 1983
6. Salmon SE, Alberts DS, Meyskens FL, Durie BGM, Jones SE, Soehnlein B, Young L, Chen GH, Moon TE: Clinical correlation of in-vitro drug sensitivity. In: Salmon SE (ed) Cloning of human tumor stem cells. New York: Alan Liss Inc 1980: 223-246
7. Shorthouse AJ, Peckham MJ, Smyth JF, Steel GG: The therapeutic response of bronchial carcinoma xenografts: A direct patients-xenograft comparison. Br J Cancer 1980 (41, Suppl. IV): 142-145
8. Valet G, Warnecke HH, Kahle H: New possibilities of cytostatic drug testing on patient tumor cells by flow cytometry. Blut 1984 (49): 37-43
9. Volm M, Wayss K, Mattern J, Kleckow M, Vogt-Moykopf I: Resistenztestung und Chemotherapieergebnis bei Bronchialtumoren. Dtsch med Wschr 1978 (103): 1266-1270
10. Von Hoff DD, Casper J, Bradley E, Sandbach J, Jones D, Makuch R: Association between human tumor colony forming assay results and response of an individual patient's tumor to chemotherapy. Am J Med 1981 (70): 1027-1032
11. Weisenthal LM, Marsden JA, Dill PL, Macaluso CK: A novel dye exclusion method for testing in-vitro chemosensitivity of human tumors. Cancer Res 1983 (43): 749-757

III. Prediction of Antitumour Activity per Tumour Type by the Human Tumour Xenograft

Conventional Agents in Human Ovarian Cancer Xenografts

E. Boven

Department of Oncology, Free University Hospital, De Boelelaan 1117, 1081 HV Amsterdam, The Netherlands

Abstract

Ten human ovarian cancer lines, differing in histologic subtype and growth rate, were used to study the efficacy of 6 conventional antitumor agents. Tumor growth inhibition of more than 75% was obtained with cisplatin in 4/10 lines, with cyclophosphamide in 3/10 lines, with adriamycin in 4/10 lines, and with hexamethylmelamine in 5/10 lines. Both methotrexate and 5-fluorouracil did not show growth inhibition of importance. The chemosensitivity pattern of the panel resembled the pattern which can be expected when treating ovarian cancer patients with the same drugs.

Introduction

In general, the chemosensitivity of a human tumor line reflects the response to the same drugs obtained in the donating patient (1-4). This observation emphasizes the potential value of a panel of lines for its use as a model to predict the efficacy of new drugs in a particular human malignancy. Similarly to a phase II trial in patients, the model for drug screening should contain a series of lines of one tumor type, differing from each other in histology, growth rate and chemosensitivity. The model will only be reliable if a good correlation exists between mouse data and results in patients, both for active and inactive agents.

In a number of human tumor lines of various histologic subtypes and doubling times, we investigated the efficacy of 6 conventional cytostatic agents. All drugs have been studied as single agents in phase II trials in patients and the response rate varied from approximately 45% for cyclophospamide (CTX) and cisplatin (CDDP) to less than 20% for 5-fluorouracil (5-FU) and methotrexate (MTX) (Table 1).

Table 1. Response rates of single agents as initial therapy in patients with ovarian cancer (5)

Drug	No. of evaluable patients	No. of clinical responses	(%)
alkylating agents	2037	851	(42)
cisplatin	44	21	(48)
hexamethylmelamine	54	17	(32)
adriamycin	77	27	(35)
5-fluorouracil	24	4	(17)
methotrexate	10	1	(10)

Materials and Methods

Animals and Tumor Lines

The animals used were female 8 to 10 week old nude mice of the B10 LP/cpb strain. Ten tumor lines were studied and have been described in Chapter 2 of this Monograph. Transplantation was carried out by bilateral implantation of tumor fragments of 2-3 mm diameter in the subcutaneous region behind the front leg.

Treatment and Evaluation

Nude mice bearing tumors of 50-150 mm^3 were randomized in groups of 6-7 for treatment and 4-5 for control. Drugs were administered at the maximum tolerated dose (MTD); a weight loss up to 15% in the first week of treatment was allowed. Drugs, doses and days of administration are shown in Table 2. Tumors were measured once or twice a week and the volume was expressed in mm^3 (0.5 length × width × thickness). Relative volumes were calculated as V_T/V_0, where VT represents a volume on an individual day and V_0 the volume at the start of treatment. Growth delay was measured using the ratio of the mean of the relative volumes of treated and the mean of the relative volumes of control tumors × 100 (T/C%) and expressed as the

Table 2. Drugs, doses and schedules used in human ovarian cancer lines

Drug	Dose (mg/kg)	Route	Days of treatment
cisplatin	4–5	i.v.	0, 7 (, 14)
cyclophosphamide	150	i.p.	0, 14
adriamycin	8–10	i.v.	0, 7
hexamethylmelamine	150	i.p.	0, 2, 4, 6
methotrexate	150	i.p.	0, 7 (, 14)
5-fluorouracil	60	i.p.	0, 7, 14, 21

Table 3. Inhibition of tumor growth (%)[a] by single agents in human ovarian cancer xenografts

Tumor line	CDDP(%)	CTX(%)	ADM(%)	HMM(%)	MTX(%)	5-FU(%)
Ov. Pe	+ (55)	– (36)	+ (52)	+ (60)	– (30)	– (19)
MRI-H-207	++ (100)	++ (100)	++ (99)	++ (100)	+ (75)	– (37)
Ov. Me	+ (64)	++ (98)	++ (95)	++ (99)	+ (52)	– (14)
Ov. He	+ (56)	– (10)	+ (66)	+ (73)	– (7)	– (43)
Ov. Gl	++ (77)	– (44)	– (44)	++ (87)	o	– (14)
Ov. Rı(C)	++ (92)	+ (62)	– (35)	+ (61)	o	– (26)
FCo	– (46)	– (5)	– (0)	– (36)	o	– (0)
FKo	– (14)	– (4)	– (8)	– (35)	o	– (21)
FMa	++ (99)	+ (60)	++ (76)	++ (98)	– (0)	+ (54)
A 2780S	+ (64)	++ (82)	++ (81)	++ (80)	o	– (4)

[a] growth inhibition ≤50% –; >50% to ≤75% +, >75% ++; o not done.

percentage of growth inhibition (100% – T/C%). The lowest value within 5 weeks after the last injection was the optimal T/C%. Inhibition of growth >75% was graded as ++, >50 and ≤75% as +, and ≤50% as –. A complete remission (growth inhibition of 100%) was defined as the total disappearance of all tumors for a period of at least one month. Animals dying from toxicity within 2 weeks after completing chemotherapy were considered as toxic deaths and were excluded from the study.

Results

Most drugs were administered once a week. Hexamethylmelamine (HMM) is usually given orally to patients on a daily basis, therefore alternate days were chosen for injection. Adriamycin (ADM) and cisplatin (CDDP) were injected i.v., which may offer a better distribution of these drugs as compared to the i.p. route. The efficacy obtained with the 6 drugs is summarized in Table 3. For CDDP a growth inhibition >75% was obtained in 4/10 lines, for CTX in 3/10, for ADM in 4/10 and for HMM in 5/10 lines. Both MTX and 5-FU did not show growth inhibition of importance. Except for HMM the chemosensitivity pattern resembled that which would be expected in ovarian cancer patients (Table 1).
Characteristically, the tumor lines showed an individual pattern of responsiveness to each of the agents with growth inhibition ranging from 0% to 100%.

Discussion

Ten human ovarian cancer lines, which are different in morphologic and biologic properties, were also shown to differ in their specific responsiveness to 6 conventional antitumor agents in use for the treatment of ovarian cancer. More important, the chemosensitivity pattern of the panel resembled the pattern obtained with similar drugs in the clinical situation.
Thus far, a positive correlation has been observed for antitumor effects of both active and inactive drugs in our xenografts and in human ovarian cancer patients. However, HMM showed a striking efficacy in the ovarian cancer model. Several facts might explain the discrepancy with the results of HMM in humans. Due to the limited water solubility of HMM it is administered orally in the clinic, resulting in a variable bioavailability. The metabolism of HMM is different in mouse and man, mice can generate more of the active metabolites, N-methylolmelamines (6). Incidentally, other drugs, such as MTX in head and neck cancer xenografts, have been reported to yield false-negative data (7). These observations warrant the careful interpretation of results of screening new drugs in a panel of human tumor lines.
In our human ovarian cancer model we plan to incorporate at least 14 lines, analogous to the number of patients required in a phase II clinical trial (8). By retrospective and prospective comparison of the activity of new antitumor agents in the panel of ovarian cancer xenografts and in phase II trials in patients, the reliability of this model will be evaluated for its potential contribution in improving treatment results in ovarian cancer.

References

1. Shorthouse AJ, Peckham MJ, Smyth JF and Steel GG: The therapeutic response of bronchial carcinoma xenografts: a direct patient-xenograft comparison. Br J Cancer 1983 (41) suppl. IV: 142–145
2. Fodstad Ø, Aass N and Pihl A: Response to chemotherapy of human malignant melanoma xenografts in athymic, nude mice. Int J Cancer 1980 (25): 453–458
3. Bailey MJ, Gazet JC, Smith IE and Steel GG: Chemotherapy of human breast-carcinoma xenografts. Br J Cancer 1980 (42): 530–536

4. Giovanella BC, Stehlin JS, Shepard RC and Williams LJ: Correlation between response to chemotherapy of human tumors in patients and in nude mice. Cancer 1983 (52): 1146-1152
5. Neijt JP: Combination chemotherapy in the treatment of advanced ovarian carcinoma. Dordrecht: ICG Printing 1983
6. Rutty CJ, Newell DR, Muindi JRF and Harrap KR: The comparative pharmacokinetics of pentamethylmelamine in man, rat and mouse. Cancer Chemother Pharmacol 1982 (8): 105-111
7. Braakhuis BJM, Leyva A, Schoevers EJ, Boerrigter GH, Schornagel JH and Snow GB: Lack of effect of methotrexate on human head and neck tumours transplanted in athymic nude mice. Acta Otolaryngol (Stockh) 1985 (99): 208-213
8. Bellet RE, Danna V, Mastrangelo MJ and Berd O: Evaluation of a "nude" mouse-human tumor panel as a predictive secondary screen for cancer chemotherapeutic agents. J Natl Cancer Inst 1979 (63): 1185-1188

Activity of Conventional Drugs in Head and Neck Cancer Xenografts

B. J. M. Braakhuis and G. B. Snow

Department of Otolaryngology, Free University Hospital, De Boelelaan 1117, 1081 HV Amsterdam, The Netherlands

Abstract

Nude mice bearing human head and neck cancer xenografts were treated with 5 anticancer drugs that are active in patients with this type of tumour. Using maximum tolerated dose levels, bleomycin appeared to be the most active drug, with activity in 4/9 lines tested; a growth delay factor of more than 2 (GDF > 2) was taken as criterion of activity. 5-Fluorouracil and cyclophosphamide gave only moderate responses (1 < GDF < 2) in 4/7 and 4/6 of the lines, respectively. Methotrexate and cisplatin, tested in 11 and 10 lines, respectively, were minimally or not active in the lines investigated. The activity of cisplatin appeared to be dose dependent. In the past higher doses could be given, resulting in 5/6 responding lines. In 4 lines a direct comparison could be made between the two schedules: lines that did not respond to treatment with the present schedule did so in the past, when the higher total dose was used. The activity of methotrexate in the nude mouse is less than can be expected from clinical studies. The mechanism responsible for this lack of activity is discussed in relation to our recent results.

Introduction

Human tumours can be grown in athymic mice due to their immunodeficiency. In head and neck cancer we observed a 'take rate' in the first passage of 27% (1), which is in agreement with the results of other authors (2-4). Xenografted head and neck tumours can be passaged serially in about half of these cases (1, 4).

The nude mouse model might play an important role in the screening of new agents for head and neck cancer. First, however, it has to be established how chemotherapy-results in this model correlate with the results in the clinic. This can be achieved by testing a number of conventional drugs in a panel of head and neck tumour lines. The results will also give information on the variation in sensitivity among the different lines. This is important for the choice as to how many lines have to be applied when the activity of a new drug is tested. In this paper we describe the sensitivity of a number of head and neck tumour lines to 5 drugs that are active in clinical practice (5).

Materials and Methods

Animals and Tumours

Female nude mice (B10.LP/Cpb, 8-10 weeks old) were obtained from the Centraal Proefdieren Bedrijf TNO (Zeist, The Netherlands). The conditions under which the mice are kept are reported elsewhere (6). Tumour fragments were transplanted subcutaneously. The characteristics of the lines used have been described earlier (1, 6-8). With the exception of the line HNX-DU, which originated from an undifferentiated carcinoma, all tumours were derived from squamous cell carcinomas, varying in degree of differentiation and site of origin.

Chemotherapy

Tumour volume was calculated according to the formula 0.5 × length × width × height. Animals bearing tumours with a volume between 50 and 150 mm^3 were randomly divided into treatment and control groups (5-8 tumours/group). Drugs were given intraperitoneally at a maximum tolerated dose level (5-15% weight loss). Doses and schedules are summarized in Table 1. An antitumour effect was expressed as tumour growth delay (8). The growth delay factor (GDF) was defined as the difference between the mean values of the time required by tumours of treated and control animals to double their volume, divided by the mean value of the time required by the tumours of the control mice to double their volume.

Table 1. Efficacy of conventional drugs in human head and neck cancer xenografts

Drug	Dose (mg/kg)	Schedule	GDF[a] >1	GDF >2
Methotrexate	250	q7d×2	1/11[b] (9%)	0/11 (0%)
Methotrexate	5	q1d×5	1/10 (10%)	0/10 (0%)
Cisplatin	2	q1d×3	0/ 8 (0%)	0/ 8 (0%)
Bleomycin	15	q1d×3/4	6/ 9 (67%)	4/ 9 (44%)
5-Fluorouracil	25	q1d×4	4/ 7 (57%)	0/ 7 (0%)
Cyclophosphamide	100	q7d×2	4/ 6 (67%)	0/ 6 (0%)

[a] growth delay factor; see materials and methods.
[b] no. sensitive/no. tested lines.

Table 2. Dose-dependent effect of cisplatin on response of xenografts

Line	Early passage			Late passage		
	No.	Total[a] dose	GDF	No.	Total[a] dose	GDF
HNX-FR	5	9	1.0	8	6	0.1
HNX-E	6	9	1.4	9	6	0.2
HNX-KE	6	9	1.5	10	6	0.5
HNX-G	6	12	1.4	11[b]	6	0.1

[a] in mg/kg; for details, see Table 3.
[b] after freezing and thawing.

Table 3. Response to cisplatin in head and neck cancer xenografts

Total dose (mg/kg)	Dose (mg/kg)	Schedule	Sensitivity criterion	No. sensitive/ no. tested lines
9–15	3	q1d×3–5	GDF>1	12/13 (92%)
	3	q1d×3–5	GDF>2	6/13 (46%)
6	2	q1d×3	GDF>1	0/ 8 (0%)

[a] growth delay factor: see materials and methods.

Results and Discussion

Five conventional agents that are active in the clinic in 20–40% of patients (5) were tested on human head and neck cancer xenografts (Table 1). Two criteria of activity were chosen: drugs causing a GDF of more than 1 and less than 2 were defined as moderately active while in active drugs the GDF has to be more than 2. With the dose schedules used bleomycin appeared to be the most active drug (4/9 lines). 5-Fluorouracil and cyclophosphamide were only moderately active in 4/7 and 4/6 of the lines, respectively. Methotrexate (MTX) and cisplatin (CDDP) were minimally or not active.

The poor activity of CDDP reported here is not in accordance with our previously reported results (8), where 5 out of 6 lines showed a GDF>2. The experimental set-up could explain the difference in activity. In our earlier studies a higher total dose could be given than used here. At this moment we are forced to use lower doses of CDDP. The same strain of mice is used, but due to some unexplained cause the mice tolerate less.

In 4 lines this dose-dependent response was also observed in a direct retrospective way. Treatment with the low total dose of 6 mg/kg CDDP always resulted in a GDF of less than 0.5, whereas higher total doses caused a GDF of more than 1 in all lines (Table 2).

Summarizing the results of all our experiments with CDDP, we came to the conclusion that a higher total dose of CDDP leads to a response in a larger proportion of lines tested (Table 3).

Also Wennerberg et al. (9) observed a dose-dependent effect of CDDP in human head and neck cancer xenografts. Interestingly, with bleomycin we did not observe such a strong dose-response relationship. The overall sensitivity to this drug did not change much when the total dose was decreased. The fact that the dose-dependent responsiveness varies with the drugs chosen can have implications for the screening of new drugs. In our case CDDP would only have been detected as an attractive drug in our earlier studies in which the mice could tolerate more of the drug. This can also happen when two strains of mice are used with a different CDDP tolerance. Thus, the choice of the strain of mice may influence the response to CDDP and perhaps also the response to new drugs.

We have tested MTX with a daily dose schedule in 10 lines and only in 1 line a moderate response was observed. The same applies for the weekly dose schedule. As was reported previously (8), in 2 cases a direct comparison could be made between the responses of the patients' tumours and the tumour lines derived from these tumours. The patients' tumours regressed after treatment whereas the corresponding xenografts did not respond to treatment, irrespective of the schedule used (7, 8). In our opinion the study of the underlying mechanisms of this insensitivity is very important. What should be determined is the effect of the host on the tumours when they are grafted. This study might also reveal resistance mechanisms, additional to the ones known from normal mice and *in vitro* studies.

MTX is an inhibitor of dihydrofolate-reductase (10). This enzyme is responsible for the conversion of dihydrofolate to reduced folate factors. This leads to an inhibition of the synthesis of purine and pyrimidine nucleotides, precursors of DNA and RNA. Also one or more glutamates can be chemically bound to MTX to form MTX-polyglutamates, which are retained longer in the cell (11).

To date we have investigated several possible mechanisms in order to explain this insensitivity to MTX in head and neck tumour lines. It is unlikely that the lack of effect of MTX in head and neck cancer xenografts is due to a difference in plasma pharmacokinetics between man and nude mice, since a xenografted rat tumour was found to be sensitive (7). Furthermore, we did not find evidence for increased dihydrofolate-reductase activity in the MTX-insensitive xenografts (7). There is also evidence that impaired drug transport does not play an important role in this insensitivity, since MTX is taken up by the cells measured as an inhibition of DNA-synthesis in insensitive tumour lines after MTX administration (B.J.M. Braakhuis et al., manuscript in preparation). A possibility is that the retention of MTX is too short, and this is perhaps caused by a decreased synthesis of MTX-polyglutamates.

In conclusion, our results show that with some drugs problems arise in the nude mouse xenograft model. The activity of MTX is less than expected from clinical studies and it is worth noting that this also applies to another anti-metabolite, 5-fluorouracil, when tested on gastrointestinal tumour xenografts (12). For CDDP a dose-dependent effect was observed, which can have implications for the choice of strain of mice. Due to the limited number of lines tested, it is at this moment too early to draw conclusions regarding a correlation between the xenograft model and the clinical situation.

References

1. Braakhuis BJM, Sneeuwloper G and Snow GB: The potential of the nude mouse xenograft model for the study of head and neck cancer. Arch Otorhinolaryngol 1984 (239): 69-79
2. Lindenberger J, Ganzer U and Fortmeyer HP: Die Heterotransplantation bösartiger Kopf- und Halsgeschwülste auf thymusaplastische nackte Mäuse. Arch Otorhinolaryngol 1978 (220): 117-128
3. Wennerberg J, Tropé C and Biörklund A: Heterotransplantation of human head and neck tumours into nude mice. Acta Otolaryngol 1983 (95): 183-190
4. Hill JH, Plant RL, Harris DM, Grossweiner LI, Rok B and Seter AJ: The nude mouse xenograft system: a model for photodetection and photodynamic therapy in head and neck squamous cell carcinoma. Am J Otolaryngol 1986 (7): 17-27
5. Hong WK and Bromer R: Chemotherapy in head and neck cancer. N Engl J Med 1983 (308): 75-79
6. Braakhuis BJM, Nauta MM, Romijn JC, Rutgers DH and Smink T: Enhanced success rate of transplantation with human tumors in cyclophosphamide-treated nude mice. J Natl Cancer Inst 1986 (76): 241-245
7. Braakhuis BJM, Leyva A, Schoevers EJ, Boerrigter GH, Schornagel JL and Snow GB: Lack of effect of methotrexate on human head and neck tumours transplanted in athymic nude mice. Acta Otolaryngol 1985 (99): 208-213
8. Braakhuis BJM, Schoevers EJ, Heinerman ECM, Sneeuwloper G and Snow GB: Chemotherapy of human head and neck cancer xenografts with three clinically active drugs: cisplatinum, bleomycin and methotrexate. Br J Cancer 1983 (48): 711-716
9. Wennerberg J, Alm P, Biörklund A, Killander D, Långström E and Tropé C: Cell cycle perturbations in heterotransplanted squamous cell carcinoma of the head and neck after mitomycin c and cisplatin treatment. Int J Cancer 1984 (33): 213-222
10. Schornagel JH and McVie JG: The clinical pharmacology of methotrexate. Cancer Treat Rev 1983 (10): 53-75
11. Jolivet J, Schilsky RL, Bailey BD, Drake JC and Chabner BA: Synthesis, retention and biological activity of methotrexate polyglutamates in cultured human breast cancer cells. J Clin Invest 1982 (70): 351-360
12. Fiebig HH, Schuchhardt C, Henss H, Fiedler L and Löhr GW: Comparison of tumor response in nude mice and in the patients. Behring Inst Mitt 1984 (74): 343-352

Preclinical Drug Evaluation with Human Melanoma Xenografts

R. OSIEKA

Innere Universitätsklinik und Poliklinik, West-German Cancer Center, University of Essen, Hufelandstraße 55, 4300 Essen, Federal Republic of Germany

Introduction

In 1974 there was an increasing awareness at the National Cancer Institute of the U.S.A. that important drugs might have been missed by the conventional murine tumor screening system then employed. Differences in biochemistry and cell biology of murine vs. human tumors were increasingly well understood. Among these were growth kinetic properties, membrane transport rates, preferential metabolic pathways, nucleotide pool regulatory mechanisms, and preferential mode of DNA repair.

Under the direction of Dr. A. Goldin this problem was approached by comparing drug responses of murine colon cancer vs. human colon cancer xenografts. Drugs with proven efficacy in conventional organ-specific murine systems, e.g., transplantable colon cancer line Co 38 (1), underwent testing against a panel of 3 human colon cancer xenografts of different histologic grade (2–5).

Table 1 gives a comprehensive though not complete overview of the responses observed in the colon cancer xenograft model (5). A quick glance illustrates the sensitivity of the new screening tool: most agents were rejected as ineffective. A much more detailed presentation of results in experimental chemotherapy of colonic tumors of murine vs. human origin is given by Venditti et al. (6).

From an ethical point of view it is deplorable that these negative results were not heeded immediately. The negative preclinical results were later confirmed by clinical trials (7). Thus, AMSA, anguidine, maytansine and PALA were rightly rejected by + his new model. At second look, however, the Table reveals some problems despite its seemingly excellent predictions. With respect to the degree of histologic differentiation the whole range of possibilities was covered, but as far as differential response to drugs was concerned this was only achieved for methyl-CCNU. Complete tumor regressions were achieved in colon line CX-3 alias 'BE' vs. no response in line CX-1 alias 'HT-29' and partial response in line CX-2 alias 'CA'. A crucial line of investigations originated from the differential response of these tumor lines to methyl-CCNU which ultimately led to the identification of the repair-efficient Mer^+ vs. the Mer^- phenotype (8, 9).

For many other agents the screen CX-1, CX-2 and CX-3 did not cover the whole range of responses. For example, the human colon cancer line LoVo is highly responsive to cisplatin (10) but no tumors responding to cisplatin were included in our panel. This statement is somewhat in conflict with a previous publication (6) indicating activity of cisplatin against human colon xenografts CX-1 sufficiently high to warrant presentation to decision network of the NCI. A careful analysis of dose responses up to 9 mg/kg of cisplatin failed to confirm significant activity of this drug (R. Osieka, unpublished observations).

Table 1. Response of 3 human colon cancer lines to agents active against the murine tumor Co38

Drug	Dose (mg/kg)	Schedule[a]	Response of line		
			CA	BE	HT
PALA	500	q4d × 3 i.p.	–	–	–
AMSA	4.8, 8.0, 13.3	q4d × 3 i.p.	–	–	–
Maytansine	0.108, 0.180	q4d × 3 i.p.	–	–	–
Lymphosarcin	3.6, 6.0, 10.0	q4d × 3 i.p.	–	–	–
Baker's antifol	25, 50, 100	q4d × 3 i.p.	–	+/–	–
Anguidine	5.6	q4d × 3 i.v.	–	–	–
Methyl-CCNU	2.3–30.0	s.d. i.p.	+/–	+	–
Chlorozotocin	12.0, 20.0	s.d. i.p.	–	+	–
Cyclophosphamide	90, 150	s.d. i.p.	–	–	–

[a] q4d × 3, Three times every 4 days; i.p., intraperitoneally; i.v., intravenously; s.d., single dose.

Rationales of the Melanoma Xenograft Model

Several considerations led us to switch from the small panel of colon cancer xenografts (CX-1, CX-2, and CX-3) to malignant melanoma xenografts as a model for preclinical drug evaluation.

- Heterogeneity of response is observed in cancer patients according to age, primary lesion and tumor burden. Statistics therefore rule that, in order to reject a new drug, it should fail in at least 14 different patients in a disease-oriented phase II study. The same should apply to human tumor xenografts.
- The influence of the site of visceral tissue transplantation was usually neglected by the subcutaneous assay but this would not apply to malignant melanoma. This is not to say that the subcutaneous melanoma xenografts bear much analogy to the complex histological evolution from a nevus to a malignant melanoma.
- Sequential biopsies should be available.
- In order to validate the responses in xenografts there should be no ethical restrictions for single agent treatment in the clinic.
- Since there were considerable logistical restrictions as to the number of experiments to be performed, the range of compounds to be tested had to be narrowed down to only one class of agents. Drugs to be evaluated were limited to agents that interact with DNA by forming some kind of firm chemical bond such as the alkylating agents or the heavy metal compounds.

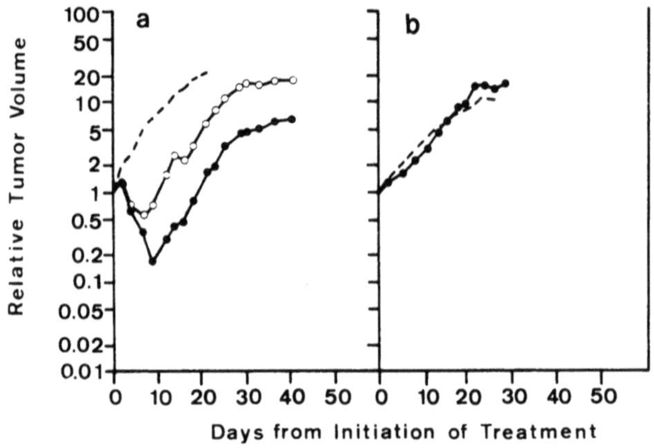

Fig. 1a, b. Tumor volume response of human melanoma xenografts in nude mice. Panel **a:** Line 'Str'; dashed line: untreated controls, open circles: 1.5 mg/kg DTIC; closed circles: 3.1 mg/kg DTIC. Panel **b:** Line 'GrII'; dashed line: untreated controls; closed circles: 200 mg/kg DTIC. Each treatment group consisted of 6-9 mice receiving treatment i.p. at day 0. Tumor volumes were set to 1 at this time

Response of Melanoma Xenografts to Drug Treatment

Malignant melanoma was the disease to fulfill most of the outlined criteria and we proceeded to establish such a screen. Donor patients received treatment with dacarbazine (DTIC) or ifosfamide together with cisplatin.
The response patterns of such a xenograft screen had to be most likely rather trivial if it were to reflect the direful results of chemotherapy in human melanoma patients. Drug-responsive tumors are therefore quite interesting with respect to detailed analysis of dose responses. Figure 1 shows the extremes of the melanoma xenograft panel with respect to drug responses: Xenograft line 'Str' is exquisitely sensitive to DTIC and the donor patient achieved a complete remission whereas xenograft line 'GrII' was derived from a patient refractory to all chemotherapy and never responded to any experimental treatment. The details of transplantation, tumor volume measurements and calculation of specific growth delay values have been reported previously (9, 12).
Table 2 displays our overall results in the treatment of xenografts. We included 'monofunctional' and 'bifunctional' alkylating agents and 2 platinum compounds presumably not cross-resistant. The data on preclinical vs. clinical responses have been reported previously and are summarized here. Suffice it to say that a growth delay value less than 2 was found to predict well for a clinical tumor progression despite adequate drug treatment. It remains an intriguing question if the antitumor potential apparent from these studies (about ¾ of all xenograft lines respond to at least 1 drug with a growth delay factor >2) would have been of therapeutic value to the patients.

Cross-Resistance Patterns

Limiting the experimental design to melanoma xenografts and a battery of quasi-alkylating agents enabled us to address a crucial question of cross-resistance patterns.
Our data show quite clearly that cross-resistance among these agents is incomplete within a wide range of sensitivities. Each xenograft line displays a unique pattern of sensitivity but some of these patterns do overlap. Even if a given xenograft line is highly responsive to several drugs, dose response curves reveal remarkable differences in sensitivity.

Table 2. Growth delay factor vs. clinical responses in human melanoma

Tumor	TD[a]	DDP[b]	DBD[b]	DTIC[b]	IF[b]	MeCCNU[b]	MMC[b]	PHM[b]
GR II	4.1	−0.2	−0.4	−0.3	−0.1	−0.2	−0.4	–
		R[c]	R	R	R	–	–	–
Ja I	4.0	1.3	5.1	−0.2	2.6	0.8	2.1	1.5
		S	–	–	S	–	–	–
Ja II	2.8	1.0	–	–	1.5	–	0.9	–
		R	–	–	R	–	–	–
Ka	8.9	0.3	0.2	1.3	0.1	0.8	4.5	−0.1
		R	–	R	R	–	–	–
Ki	2.5	5.1	0.7	1.0	4.4	2.5	1.7	0.5
		S	–	R	S	–	–	–
Ku	6.2	1.5	0.8	0.6	0.5	3.9	0.9	1.0
		R	–	R	R	–	–	–
Mo	3.4	4.1	0.1	−0.1	0.2	−0.1	−0.1	−0.1
		R	–	R	R	–	–	–
Ni	9.5	–	–	>2[d]	0.8	–	–	–
		–	–	R	–	–	–	–
Str	4.1	2.4	0.9	>2	>2	>2	>2	0.1
		S	–	S	S	–	–	–
St	3.6	–	–	>2	–	–	–	–
		R	–	S	R	–	–	–
Zi	12.1	0.2	–	0.1	0.4	–	–	–
		R	–	R	R	–	–	–
Kr I	8.1	−0.4	–	>0.5	−0.2	2.9	3.1	–
		R	–	R	R	–	–	–
Flo	7.6	0.7	–	−0.2	>2	−0.2	1.1	–
		–	–	R	–	–	–	–
Avo	9.1	0.8	–	0.1	–	0.5	–	–
		–	–	R	–	–	–	–
Da I	19.8	−0.4	–	>2	−0.4	–	–	–
		R	–	R	R	–	–	–
De II	14.5	>2	–	>2	>2	>2	–	–
		–	–	S	S	S	-	–
Alt	9.4	–	–	−0.1	1.4	>2	–	–
		–	–	R	–	–	–	–
Sch	3.1	1.7	0.3	1.28	>2	>2	>2	0.2
		–	–	R	–	–	–	–
Ge	16.9	–	–	>2	–	–	–	–
		–	–	–	–	–	–	–
Re	8.9	–	–	−0.3	0.8	–	–	–
		–	–	R	–	–	–	–

[a] TD = Tumor volume doubling time for untreated animals (days).
[b] DDP = cisplatin, DBD = dibromodulcitol, DTIC = dacarbazine, IF = ifosfamide, MeCCNU = methylCCNU, MMC = mitomycin C, PHM = malonato-DACH-platinum II.
[c] clinical evaluation: R = resistant; S = sensitive; – = not evaluated.
[d] Values >2 were taken for growth delay, if doubling of tumor volume was not reached within 60 days.

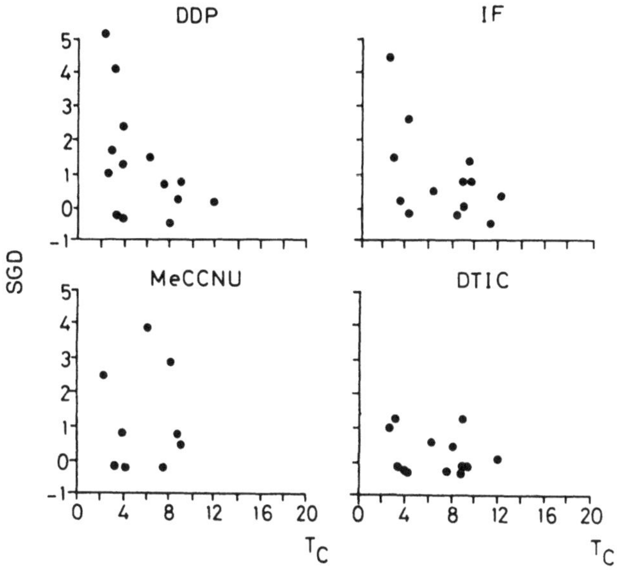

Fig. 2. Specific growth delay values (SGD) are plotted vs. tumor volume doubling times (T_C) for each drug. Abbreviations are: DDP = cisplatin, (N = 15, rho = −0.508); MeCCNU = methyl-CCNU (N = 13, rho = 0.172); IF = ifosfamide, (N = 17, rho = −0.315); DTIC = dacarbazine, (N = 19, rho = 0.28)

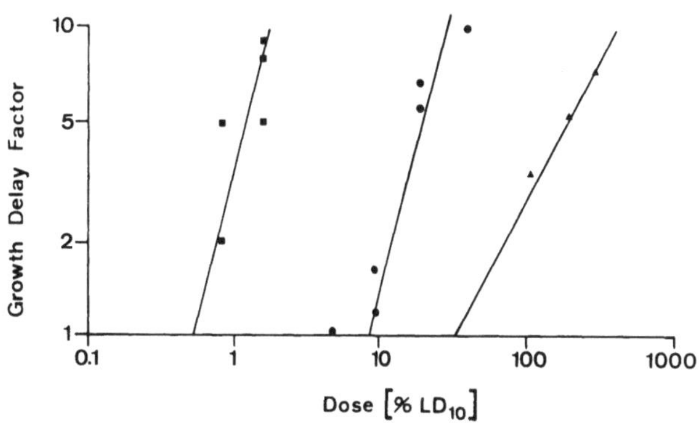

Fig. 3. Dose-response curves for dacarbazine (rectangles), melphalan (circles) and methyl-CCNU (triangles) using human melanoma xenograft line 'Str'. Fractions of the $LD_{10/30}$ values in percent were used as a common denominator for dose

This refutes the notion of tumors being either globally sensitive or resistant. Plotting specific growth delay values achieved by single agent treatment against tumor volume doubling times shows a poor correlation between these values (Fig. 2).

For the purpose of ranking the activity of various drugs against a single tumor line, dose response curves were constructed with fractions of $LD_{10/30}$ values as the common denominator. An almost hundredfold difference in sensitivity becomes apparent from the position of the dose-response curves for 1 xenograft line that might have been judged as globally sensitive if challenged with each drug at the $LD_{10/30}$ level only.

Selection of Resistant Xenograft Sublines

Resistant sublines should not only be developed by *in vitro* exposure but also by *in vivo* exposure, since tumorogenicity and metastatic potential may be lost under *in vitro* selection pressures.

A challenge of the resistant sublines with non-selecting agents again shows incomplete cross-resistance (Fig. 4). There is a hint that resistance to bifunctional alkylating agents is undirectional with monofunctional agents (11).

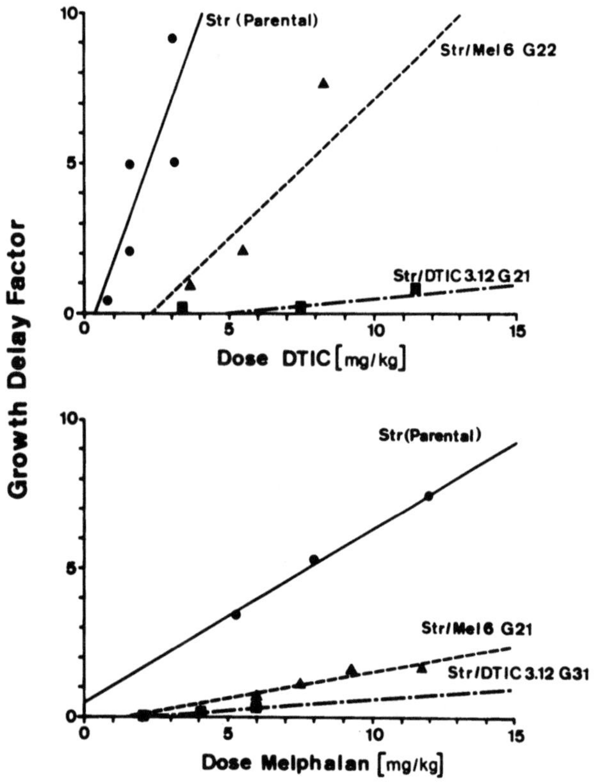

Fig. 4. Dose-response curves for parental human melanoma xenograft 'Str' vs. sublines resistant to melphalan ('Str/mel 6 G22') and dacarbazine ('Str/DTIC G21' and 'Str/DTIC G31'). The figure behind G denotes the number of tumor passages under treatment. The change in slope of the dose-response curve for dacarbazine is more pronounced at G 21 than the change in slope of the dose-response curve for melphalan considering the sublines 'Str/DTIC'. In contrast to this, the change in slope of the dose-response curves is not much different if subline 'Str/Mel' is challenged with melphalan or with DTIC

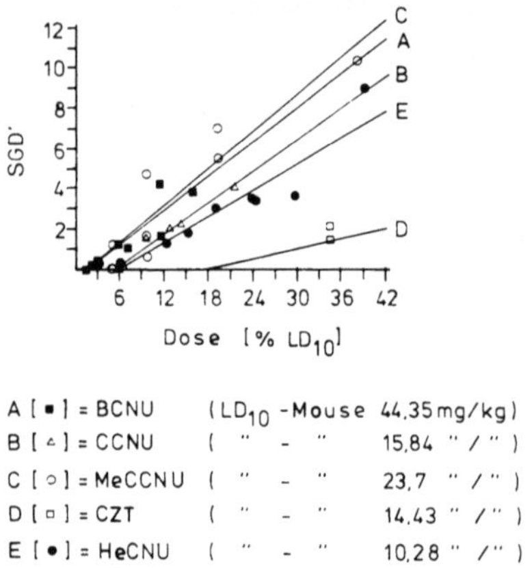

Fig. 5. Specific growth delay (SGD) as a function of dose expressed as fractions of the respective $LD_{10/30}$ values in percent. The water-soluble nitrosoureas (CZT = chlorozotocin and HeCNU = Hydroxyethyl-CNU) have a more shallow dose-response curve than the lipophilic nitrosoureas (BCNU, CCNU and MeCCNU)

The data shown in Table 2 also indicate that lack of cross-resistance in one tumor type does not apply to another tumor type. The malonato-diamminocyclohexane substituted platinum derivative (PHM) was quite effective against an Ehrlich Ascites Tumor-tumor line resistant to cisplatin (13, 14) but failed to display any significant activity against subcutaneously transplanted melanoma xenografts.

Evaluation of Analogs with the Xenograft Model

The impressive responses of the human melanoma xenograft line 'Str' to methyl-CCNU prompted us to start a comparative investigation on 5 nitrosourea derivatives.

Clinical trials comparing different nitrosoureas are often hampered by the fact that responses to drug treatment are too rare to allow for statistical comparisons. This was circumvented by using a nitrosourea-sensitive melanoma xenograft.

There exists a long-standing hypothesis that myelotoxicity is reduced in water-soluble nitrosoureas. When Dr. Eisenbrand made hydroxyethyl-nitrosourea available through the German Drug Development Progam we started a comparison of 2 water-soluble and 3 lipophilic nitrosoureas in the nitrosourea-sensitive melanoma line 'Str'. There was a difference in potency among these drugs with hydroxyethyl-nitrosourea ranking first. Unfortunately, this ranking was a similar with respect to drug induced lethality. Of interest is the more shallow slope of the curve for the water soluble derivative chlorozotocin. One might speculate that this shallow dose-response in lethality experiments led investigators to underestimate the toxicity of this compound. While replotting the dose-response curves with dose expressed on a molar basis did not change the ranking, the water-soluble nitrosoureas fell to last place if doses were expressed as fractions of the $LD_{10/30}$ values of each drug (Fig. 5).

This therapeutic ranking is not changed much if human MTD values are substituted for murine lethality data. Thus, the xenograft model combined with murine toxicity data gave useful predictions of therapeutic ranking. In this context it should be mentioned that at equi-effective doses water-soluble nitrosoureas induced DNA-damage to murine bone marrow cells and reduced cellularity to a similar degree as lipophilic nitrosoureas (14). So far, clearcut clinical results with respect to therapeutic ranking of lipophilic vs. water-soluble nitrosoureas are lacking.

In Vivo Chemosensitization

Reviewing published data on attempts to improve response to nitrosoureas by chemosensitizers rather than by structural modification of the molecule, we saw some controversial results. At the Southern Research Institute, Rose and Schabel (for references see 15) had completed a very thorough investigation on the interaction of chlorpromazine and caffeine with nitrosoureas. Their data, derived from experiments using murine B 16 melanoma, had suggested a therapeutic benefit for this combination.
Later, Cohen et al. (for references see 15) conducted a clinical trial randomizing 157 patients with advanced melanoma to receive either methyl-CCNU alone or methyl-CCNU together with caffeine and chlorpromazine. There were no significant differences in response rates nor in toxicity. The authors could not give an explanation for this failure of chemosensitization but felt that the low overall response rates might have contributed to the problem.
We addressed this problem by using a human melanoma xenograft line with well defined sensitivity to methyl-CCNU and essentially repeated the treatment schedules developed by Rose and Schabel. Our investigations demonstrated a significant enhancement of antineoplastic activity by chlorpromazine and caffeine (Fig. 6). The influence on lethality was tested again and similar enhancement was found.
These investigations were completed by demonstrating increased DNA damage and reduced cellularity in murine bone cells after combined treatment (15).

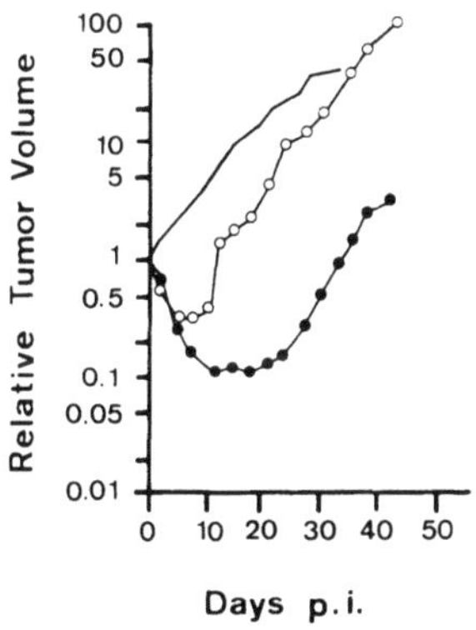

Fig. 6. Tumor volume response of human melanoma xenograft 'Str' to 3 mg/kg methyl-CCNU given i.p. alone (solid line), 3 mg/kg methyl-CCNU i.p. + 100 mg/kg caffeine s.c. (open circles), and 3 mg/kg methyl-CCNU i.p. + 100 mg/kg caffeine s.c. + 15 mg/kg i.p. chlorpromazine

Limitations of Clinical Trials

This example was chosen to illustrate the possible shortcomings of clinical drug evaluation. The importance of dose-response relationships has been repeatedly stressed by prominent clinical investigators (16), but published data to this effect have been limited to tumor entities responding well to cytotoxic treatment. For a tumor with low response rates, even a carefully designed study with high numbers of patients in each treatment arm may fail to disclose significant differences with respect to response rates or survival.
The comparison of response rates in two kinds of treatment often implies that equitoxic doses were applied. Ethical restrictions, however, call for discontinuation of treatment in the presence of progressive disease and for dose adjustments if severe toxicity occurs. Thus, unlike in experimental systems, strict adherence to projected total doses is quite often impossible in clinical trials.
Additionally, the imperative use of salvage treatments may confound the interpretation of survival data.

Characteristics of a Representative Xenograft Panel

Thus far, several researchers have been able to demonstrate good correlations between xenograft and clinical responses to drug exposure in malignant melanoma (17, 18). Useful markers as provided by cytogenetics have asserted the validity of this new tumor model (19). Some problems remain with respect to overprediction of efficacy for the nitrosoureas which may be resolved by checking repair proficiency (Mer-phenotypes).
The establishment of a panel of xenografts aims at gaining answers about drug responses hitherto only available from disease-oriented clinical phase II trials. Continuous efforts to ensure that established xenograft panels will be representative with respect to all factors known to govern response to drug treatment are therefore mandatory. Ranking of analog series and investigations of other strategies to overcome drug resistance appear to be a promising field of future applications of this advanced tumor model.
This project was supported by BDMFT and by SFB 102.

References

1. Corbett TH, Griswold PD Jr, Roberts BJ, Peckham JC, Schabel F: Evaluation of single agents and combinations of chemotherapeutic agents in mouse colon carcinomas. Cancer 1977 (40): 2660-2680
2. Johnson RK, Houchens DP: The use of human tumor xenografts for selection of new agents against tumors with response to established drugs. In: Muggia F, Rozencweig M (eds) Lung Cancer: Progress in therapeutic research. Raven Press, New York, 1979 pp 37-44
3. Osieka R, Houchens DP, Goldin A, Johnson RK: Chemotherapy of human colon cancer xenografts in athymic nude mice. Cancer 1977 (40): 2640-2650
4. Osieka R: Chemotherapy studies with human colon cancer xenografts in nude mice. In: Sigenthaler W, Luethy R (eds) Current Chemotherapy Proceedings of the 10th International Congress of Chemotherapy. Zurich, Switzerland, Sept 18-23 1977 American Society for Microbiology (2): 1149-1151
5. Osieka R, Johnson RK: Evaluation of clinical agents in phase I clinical trial and earlier stages of development against xenografts of human colon carcinoma. In: Houchens DP, Ovejera AA (eds) The use of athymic (nude) mice in cancer research. Gustav Fischer Verlag, New York, Stuttgart, 1978 pp 217-223
6. Venditti JM: Preclinical drug development: rationale and methods. Semin Oncol 1981 (8): 349-361
7. Estey E, Hoth D, Simon R, Marsoni S, Leyland-Jones B, Wittes R: Therapeutic response in phase I trial of antineoplastic agents. Cancer Treat Rep 1986 (70): 1105-1115
8. Thomas CB, Osieka R, Kohn KW: DNA cross-linking by in vivo treatment with 1-(2-chloroethyl)-3-(4-methylcyclohexyl)-1-nitrosourea of sensitive and resistant human colon carcinoma xenografts in nude mice. Cancer Res 1978 (39): 2448-2454
9. Erickson LC, Osieka R, Kohn KW: Differential repair of 1-(2-chloroethyl)-3-(4-methylcyclohexyl)-1-nitrosourea-induced DNA damage in two human colon tumor cell lines. Cancer Res 1978 (38): 802-808
10. Bergerat JP, Barlogie B, Drewinko B: Effects of cis-dichlorodiammineplatinum (II) on human colon carcinoma cells in vitro. Cancer Res 1979 (39): 1334-1338
11. Osieka R: Primary and acquired resistance to antineoplastic chemotherapy. Cancer 1984 (54): 1168-1174
12. Hill JM, Loeb E, Pardue AS, Khan A, Hill NO, King JJ, Hill RW: Platinum coordination compounds in the treatment of acute leukemia and other malignant disease with particular reference to malonato 1,2-diaminocyclohexane platinum (II). J Clin Hemat and Oncol 1977 (7): 681-699
13. Seeber S, Osieka R, Schmidt CG, Achterrath W, Crooke ST: In vivo resistance towards anthracyclines, etoposide. Cancer Res 1982 (42): 4719-4725
14. Osieka R, Glatte P, Pannenbaecker R, Schmidt CG: Therapeutic evaluation of five nitrosoureas in a human melanoma xenograft system. Cancer Chemother Pharmacol 1983 (11): 147-152
15. Osieka R, Glatte P, Pannenbaecker R, Schmidt CG: Enhancement of methyl-CCNU induced cytotoxicity by chorpromazine and caffeine in a human melanoma xenograft. Cancer Treat Rep 1986 (70): 1167-1171
16. Frei E, Canellos P: Dose: A critical factor in cancer chemotherapy. Am J Med 1980 (69): 585-594
17. Fodstad Ø, Aass N, Pihl A: Response to chemotherapy of human malignant melanoma xenografts in athymic mice. Int J Cancer 1980 (25): 453-458
18. Steel GG, Courtenay VD, Peckham MJ: The response to chemotherapy of a variety of human tumour xenografts. Br J Cancer 1983 (47): 1-13
19. Becher R, Osieka R, Hossfeld DK, Schmidt CG: Cytogenetic studies in human tumour xenografts. In: Bastert GBA et al (eds) Thymusaplastic nude mice and rats in clinical oncology. Gustav Fischer Verlag, Stuttgart, New York, 1981 pp 329-334

Lung Xenografts as a Predictive Screen for Antineoplastic Agents

J. MATTERN, K. WAYSS, and M. VOLM

German Cancer Research Center, Institute of Experimental Pathology, Im Neuenheimer Feld 280, 6900 Heidelberg, Federal Republic of Germany

Abstract

Inoculation of 337 human tumours of different types into nude mice resulted in tumour takes in 131 cases (39%). Tumour takes varied among the various tumour categories and the various histological types. From the 131 tumours growing in primary transplants, 80 tumours could be established as lines. In order to examine whether growth characteristics of untreated tumours in nude mice might be of prognostic significance for patient survival, the data from 171 non-small cell lung carcinoma xenografts were compared with the clinical data of the corresponding patients. Neither take rate nor establishment of a tumour line in nude mice was of prognostic value to the patient. In order to validate xenografts of human tumours as a predictive system, the antitumour activity of cyclophosphamide (CTX) and cisplatin (DDP) was tested against a panel of 14 human non-small lung tumour xenografts. The poor clinical response of lung tumours to CTX and DDP as reported in the literature is reflected in the lack of response of the tumour lines to these two agents.

Introduction

In order to establish the usefulness of the nude mouse-human tumour xenografts system as a predictive screen for anticancer agents, it is necessary to demonstrate that clinically active drugs such as cyclophosphamide (CTX) or cisplatin (DDP), two broad spectrum agents in cancer chemotherapy, retain the same level of activity against xenografts in nude mice as in patients. For this reason, we have tested the antitumour activity of CTX and DDP against a panel of 14 human non-small cell lung tumour xenografts and we have compared these results with the clinical results reported in the literature.

Materials and Methods

Nude Mice

The animals used in these experiments were 6-9 week old congenitally athymic nude NMRI-mice (own breeding or Zentralinstitut für Versuchstierzucht, Hannover). The mice were kept under standardized conventional conditions (Makrolon cages, 27 °C room temperature, 50% relative humidity, autoclaved bedding) in a separate room especially controlled against infections. An autoclaved special diet, and acidified water (pH 2-3) were given ad libitum.

Tumours

Human tumour transplants were established directly by inoculation of fresh tumour tissue from patients into nude mice in our laboratory. The tumour specimens were removed under sterile conditions, minced in sterile medium and then inoculated subcutaneously with a 1.5×50 mm needle into the right anterior flank of nude mice (0.1 ml/mouse).

Chemotherapy and Evaluation of Chemotherapeutic Effect

After the tumours had reached a mean diameter of about 8-10 mm, the tumour bearing mice were randomized into groups of 5-7 animals each, and drug treatment was started. Each drug was given as a single i.p. dose to the mouse: CTX (Asta-Werke, Bielefeld): 240 mg/kg; DDP (Bristol-Myers GmbH, Bergisch Gladbach): 10 mg/kg.

Both drugs were injected in a volume of 0.02 ml/g of body weight. The tumour growth was followed by measuring two diameters daily with calipers. The effect of the drugs was expressed as T/C ratio (mean tumour size of the treated tumours/mean tumour size of control group) $\times$ 100. The lowest value was expressed as an optimal T/C (%) for each group.

Results and Discussion

During the past years we have heterotransplanted more than 300 human tumours of different types into nude mice (Table 1). Tumour growth in the nude mouse was not observed in all tumour types. It was not possible to obtain growth in tumours of the thyroid gland, melanomas or kidney. The overall take rate was 39%. From the 131 tumours growing in primary transplants, 80 tumours could be established as lines transplanted beyond passage 3. Most of them were frozen in nitrogen and are available for further therapeutic and biological studies.

A total of 255 lung tumour samples from 213 patients were xenografted into nude mice (Table 2). The overall success rate for tumours of primary site was 45%. This agrees well with the experience of other workers (1, 2). It was found that all four main histopathological types of bronchial carcinoma could be grown as xenografts. Squamous cell carcinomas and adenocarcinomas gave a higher take rate (53/104 and 24/56) than large cell or small cell carcinomas (11/29 and 6/16) (Table 3).

It was more difficult to establish cell lines from adenocarcinomas than from other histological types. Fogh et al. (3) demonstrated that samples of metastases grow more readily than tumour samples of primary sites. This cannot be confirmed by our tumour material. Takes were higher for tumours of primary site (92/203) than for lymph node metastases (15/42) or recurrent tumours (4/10).

The majority of lung carcinoma xenografts studied in the present work showed volume doubling times between 10 and 20 days in passage 1. Similar volume doubling times have been reported by other workers in nude mice (4) and in immune-deprived mice (5). These volume doubling times are considerably shorter than those measured in man (6). There was an increase in growth rate from man to mouse (in four lung tumours it was possible to determine the doubling time in man by measuring changes in tumour size from consecutive X-rays and in the corresponding xenografts) and a further increase from the first passage to the second, falling from about 15 days to 7 days and remaining fairly stable in the subsequent passages (Fig. 1). This growth acceleration during the first few transplant

Table 1. Take rate of human tumours in nude mice with reference to tumour type

Histology	No. of attempts	Takes	(%)	Lines	(%)
Lung	255	111	(44)	70	(27)
Breast	35	10	(29)	1	(3)
Ovary	21	8	(38)	5	(24)
Thyroid gland	16	0		0	
Osteosarcoma	5	2		2	
Melanoma	3	0		0	
Kidney	2	0		0	
Total	337	131	(39)	80	(24)

Table 2. Take rate of bronchial carcinomas in nude mice with reference to site of tumour

Tumour	No. of attempts	Takes	(%)	Lines	(%)
Primary	203	92	(45)	56	(28)
Metastases	42	15	(36)	11	(26)
Recurrences	10	4	(40)	3	(30)
Total	255	111	(44)	70	(27)

Table 3. Take rate of bronchial carcinomas in nude mice with reference to histological type

Histology	No. of attempts	Takes	(%)	Lines	(%)
Squamous	104	53	(51)	39	(38)
Adeno	56	24	(43)	8	(14)
Large cell	29	11	(38)	8	(27)
Small cell	16	6	(38)	3	(19)
Miscellaneous	8	2		1	
Total	213	96	(45)	59	(28)

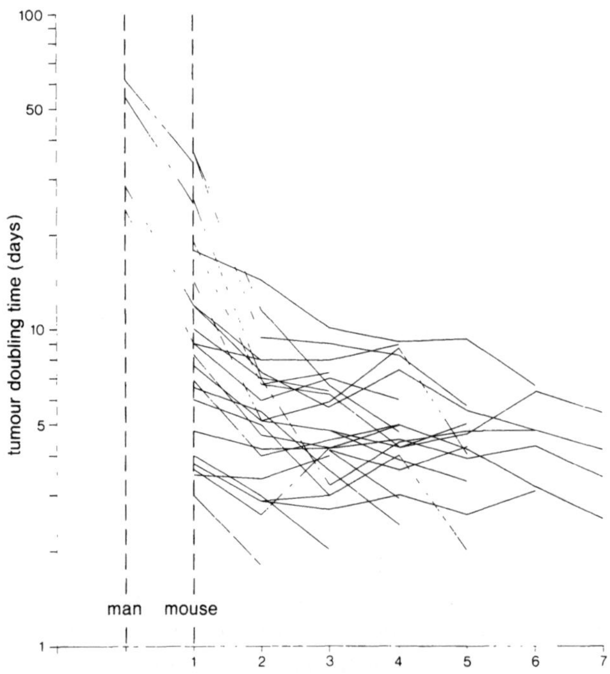

Fig. 1. Tumour doubling time of human lung tumours in man and in different passages in nude mice

generations was also observed by other workers (5, 7).

In order to examine whether growth characteristics of untreated tumours in nude mice might be of significance for patient survival, we compared the growth data of the xenografts with the clinical data of the corresponding patients. This study included 171 patients with non-small cell lung carcinomas (8). Only patients who were alive more than 4 weeks after their operation were evaluated for survival times (155). We were not able to find a relationship between tumour take in nude mice and patient prognosis nor between the establishment of a tumour line and patient prognosis (Fig. 2).

The most relevant way of validating the xenograft system for chemotherapeutic purposes is to compare directly individual xenograft response with the response of the donor patient. In the present study we also tried to make a direct patient-xenograft comparison. This was not possible since the proportion of donor patients who received chemotherapy was very small: only 3 patients from whom tumour lines could be established in nude mice were also treated with chemotherapy in the clinic. Most of the patients only underwent surgery or refused chemotherapy. Therefore, we have tested the antitumour activity of two clinically active drugs, CTX and DDP, on 14 different non-small cell lung xenograft lines in nude mice, comparing the experimental results with clinical results reported in the literature (9).

In Fig. 3, the T/C values for CTX (left) and DDP (right) were plotted versus time. It is shown that each tumour line exhibits a different response. If the results are expressed as optimal T/C values, i.e., the lowest T/C value, only 2/14 lines exhibit a value

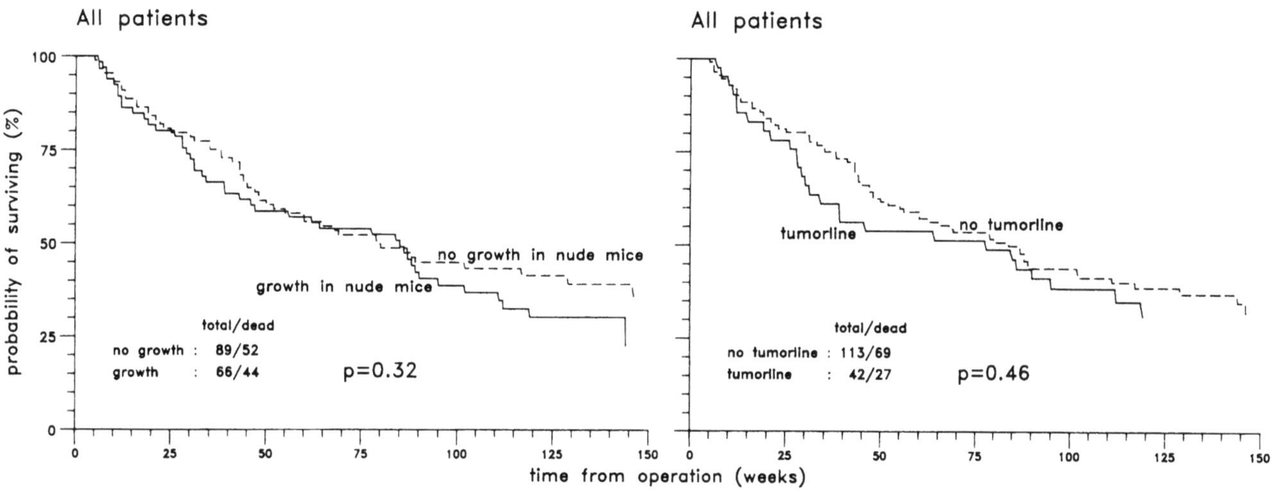

Fig. 2. Survival time of patients with non-small cell lung tumours subdivided according to the take/no take in nude mice *(left)* and to the establishment (or not) of a tumour line *(right)*

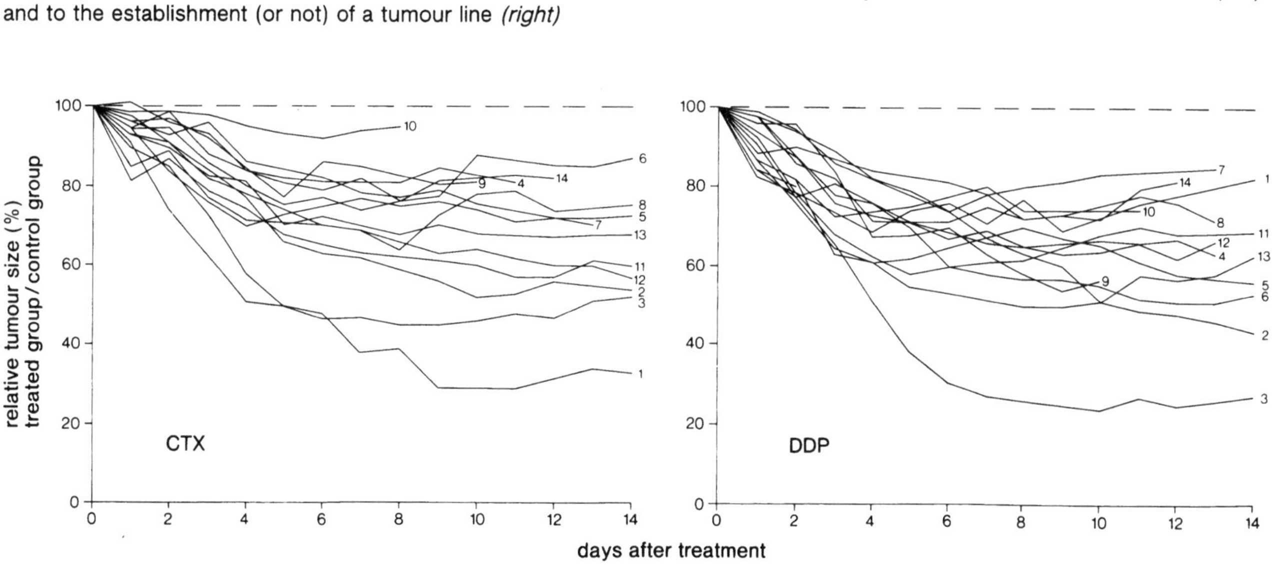

Fig. 3. Effect of a single dose of 240 mg/kg cyclophosphamide *(left)* and 10 mg/kg cisplatin *(right)* on 14 different non-small cell lung tumour lines in nude mice

Table 4. Comparison of the antitumour activity of cyclophospha mide (CTX) and cisplatin (DDP) in nude mice with clinical results reported in the literature

Drug	Xenografts	Literature	Reference
CTX	2/14 (14%)	30/369 (8%)	(9)
DDP	3/14 (21%)	28/140 (20%)	(9)

below 50% for cyclophosphamide and 3/14 lines for cisplatin.

If an optimal T/C value of less than 50% is regarded as response and the sensitivity of the lung tumour lines is compared with the clinical sensitivity of tumours of the same type as reported in the literature (9), a close correlation was found. The poor clinical response of non-small cell lung tumours was reflected by the lack of response of the tumour lines to these agents (Table 4).

Conclusions

These data support the validity of heterotransplants of human tumour in nude mice as a predictive system for new anticancer agents. The evidence for individuality of therapeutic response among the xenograft lines supports the case for incorporating panels of histologically similar xenografts into the drug-screening programme.

References

1. Sharkey FE, Fogh JM, Hajdu SI, Fitzgerald PJ and Fogh J: Experience in surgical pathology with human tumour growth in the nude mouse. In: Fogh and Giovanella (eds) The Nude Mouse in Experimental and Clinical Research 1978: 187-214 New York: Academic Press
2. Shorthouse AJ, Peckham MJ, Smyth JF and Steel GG: The therapeutic response of bronchial carcinoma xenografts: a direct patient-xenograft comparison. Br J Cancer 1980 (41) (Suppl) IV:142-145
3. Fogh J, Tiso J, Orfeo T, Sharkey FE, Daniels WP and Fogh JM: Thirty-four lines of six human tumour categories established in nude mice. J Natl Cancer Inst 1980 (61): 745-751
4. Rofstad EK, Fodstad O and Lindmo T: Growth characteristics of human melanoma xenografts. Cell Tissue Kinet 1982 (15): 545-554
5. Steel GG, Courtenay VD and Peckham MJ: The response of chemotherapy of a variety of human tumour xenografts. Br J Cancer 1983 (47): 1-13
6. Shackney SE, McCormack GW and Cuchural GJ: Growth rate patterns of solid tumours and their relation to responsiveness to therapy. Ann Intern Med 1978 (89): 107-121
7. Shimosato Y, Kameya T, Nagai K, Hirohashi S, Koide T, Hayashi H and Nomura T: Transplantation of human tumours in nude mice. J Natl Cancer Inst 1976 (56): 1251-1255
8. Mattern J, Jaeger S, Sonka J, Wayss K and Volm M: Growth of human bronchial carcinomas in nude mice. Br J Cancer 1985 (51): 195-200
9. Bakowski MT and Crouch JC: Chemotherapy of non-small cell lung cancer; a reappraisal and a look to the future. Cancer Treat Rev 1983 (10): 159-172

Chemotherapy Studies in Human Non-Small Cell Lung Cancer Xenografts Transplanted in Immune-Deprived Mice

J. CARMICHAEL and J. F. SMYTH

Imperial Cancer Research Fund, Medical Oncology Unit, Department of Clinical Oncology, Western General Hospital, Edinburgh EH4 2XU, Scotland, United Kingdom

Abstract

Non-small cell lung cancer lines were successfully established in 15/44 (34%) of tumour biopsies, 9 squamous carcinomas, 5 adenocarcinomas and 1 adenosquamous carcinoma. The overall take rate of all tumour fragments implanted in the first man to mouse passage was low (10%) but this improved in subsequent passages to 57%. Growth rates were variable in all xenografts. Histopathological features were maintained through the first 7 passages although minor changes in differentiation were noted. Chemotherapy studies using maximum tolerated doses of all drugs showed minimal single agent activity. Cis-platinum was the most active drug, exhibiting mean specific growth delay of 1.05 doubling times in 6 tumour lines, with less activity observed with 2 platinum analogues. From these growth delay results significant clinical activity would not be expected, showing correlation with clinical observations. These xenografts offer a useful model for the study of non-small cell lung cancer.

Introduction

The prognosis for the majority of patients with non-small lung cancer remains bleak. Despite recent reports of improved chemotherapeutic responsiveness in these tumours using platinum-containing regimens (1), the impact on survival of these patients has been minimal (2), with severe toxicity common (3). Suitable experimental models for the study of the biological characteristics and therapeutic responsiveness of these tumours could obviously be of great value in the development of more effective therapies.

A large number of human tumours have been satisfactorily established in a variety of immune compromised animals as reviewed by Steel in 1978 (4), with lung cancer xenografts established by a number of authors (5, 6).

In this study we report on the establishment and characterisation of human non-small cell xenografts in mice immune-deprived by means of thymectomy, cytosine arabinoside and x-irradiation as previously described (7).

Materials and Methods

Animals

The mice were supplied by the Animal Unit, Western General Hospital, Edinburgh. Neonatal mice were thymectomised at 3 weeks of age, and 2–4 weeks later received 735 cGy total body irradiation from a 250 kv source, using a Thaurus filter, giving an approximate dose of 37 cGy/minute. Cytosine arabinoside, 200 mg/kg, was administered intraperitoneally 48 hours prior to irradiation as a 'priming' dose. Following irradiation, mice were treated with terramycin and neomycin in acidified drinking water for 14 days. They were housed in a separate room in the animal facility, although specific pathogen-free conditions were not required.

Tumours

Tumour specimens were obtained primarily at thoracotomy, although a few samples were acquired following bronchoscopic investigation. The samples were collected directly from the operating theatre and transported on ice to the Pathology Department. Tumour tissue was obtained from the periphery of the primary tumour mass, where possible, and immediately placed in culture medium (Roswell Park Memorial Institute, 1640 medium with 10% foetal calf serum) at 4 °C. Specimens were immediately taken to the Animal Unit, cut into 8 mm^3 fragments and transplanted bilaterally into the flanks of 5 mice through a dorsal incision, with samples sent for histopathology to confirm the presence of tumour tissue. Transplantation was carried out within 21 days of irradiation.

Xenograft Measurement

All mice were examined twice weekly for evidence of tumour growth. When tumours became palpable they were measured twice weekly using calipers. Tumour volume was estimated from perpendicular tumour measurements using the formula for an ellipsoid:

$$\frac{\pi}{6}\,Dd^2$$

(D = longest diameter, d = shortest diameter).

Chemotherapy Studies

Treatment groups were stratified both for sex of the host and for tumour size with only tumours of greater than 0.3 cm^3 used for chemotherapy studies. Following administration of the cytotoxic drugs, tumour volume increase was compared with control tumour growth. The median tumour volume doubling time (T_D) was estimated for each treatment group, enabling the calculation of the specific growth delay (SGD) by comparison with the control group tumour volume doubling time using the formula (8):

$$SGD = \frac{T_D\,(\text{treated}) - T_D\,(\text{control})}{T_D\,(\text{control})}$$

The use of the SGD enables comparisons of chemotherapeutic activity to be made between xenograft lines with differing growth rates.

Cytotoxic Drugs

All drugs used in these studies were obtained from commercial sources, and were administered intraperitoneally at a site distal to the tumour.

Results

Xenograft Take Rates

In all, of 44 tumours specimens containing viable tumour were transplanted with xenografts being successfully established in 15 (34%). The majority of these samples, 37 (84%) were obtained at thoracotomy, the remaining 7 were bronchoscopic biopsies. Details are shown in Table 1. Biopsies obtained from rigid bronchoscopic examination resulted in the establishment of 2 squamous carcinoma xenografts. Take rates at the first man to mouse passage were low (10%), although this substantially improved on subsequent passages.

Growth Rates

There was great variability in growth rates both between xenograft lines and within the same tumour line, even with the same passage of tumour. This variability was observed both for the lag period prior to the appearance of detectable tumour and to the doubling time of measurable tumours. The lag period varied between 25-93 days on the first man to mouse passage (median 38 days) with less variability noted by the fourth passage (range 14-45 days). Median doubling times ranged from 11-20 days in the first man to mouse passage of these xenografts with marked variability noted. On subsequent passages the doubling time varied but with no specific trend, some xenografts doubling more rapidly, others more slowly. With returning immunity in some of these animals, stasis of tumour growth or even complete regression was observed, particularly in small tumours.

Histopathology

All tumours were classified according to the WHO histological classification (WHO 1981) (9) and graded using the Edinburgh classification (1982) (10). All of the xenografts maintained the general characteristics of the original tumour with variations

Table 1. Xenograft take rates of non-small cell lung cancer specimens

Histological type	No. of tumours biopsied	Established tumour lines	Take Rate[a] 1st passage	Take Rate[a] subsequent passages
Squamous	20	9 (45%)	26/180 (14%)	1408/2328 (61%)
Adenocarcinoma	16	5 (31%)	10/159 (6%)	344/ 680 (51%)
Adenosquamous	3	1 (33%)	5/ 30 (16%)	
Undifferentiated carcinoma	5	0 (0)	0/ 5 (0%)	
Total	44	15 (34%)	4/419 (9.8%)	1861/3269 (57%)

[a] No. of implants exhibiting progressive growth/total no. of implants.

in differentiation observed occasionally. Both increased and decreased differentiation was noted. Examples of histopathology at early and late passages are illustrated in Figures 1a, 1b, 2a and 2b.

Chemotherapy Studies

Six non-small cell carcinoma tumour lines were used for chemotherapy studies, 4 squamous carcinomas, 1 adenocarcinoma and 1 adenosquamous carcinoma. These lines were tested against maximum tolerated doses of a number of drugs administered as single agents as shown in Table 2. Chemotherapeutic response was measured using SGD as the end point with the results illustrated in Table 3. Minimal activity was observed for any of the drugs administered as single agents with a maximal SGD of 2 tumour volume doubling times achieved with cisplatin in the xenograft line NX002. Cisplatin was the most active single agent, although a mean SGD of only 1.09 tumour doubling times was obtained. Platinum analogues JM8 and JM40 showed some activity although they were less active than cisplatin. Ifosfamide was shown to have some activity (mean SGD 0.96), only minimal activity was observed with vindesine, VP16 or cyclophosphamide in these xenografts.

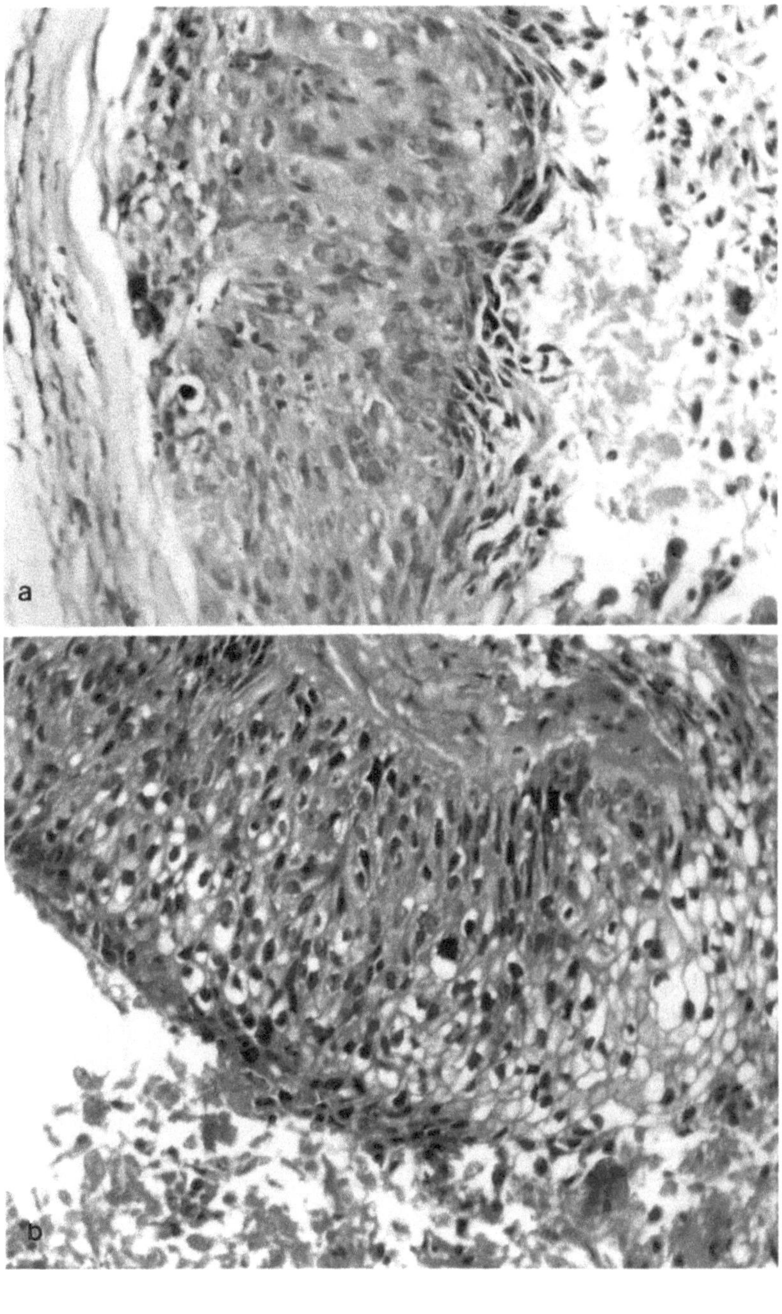

Fig. 1. a NX002 moderately differentiated squamous carcinoma - original passage. **b** NX002 7th mouse to mouse passage

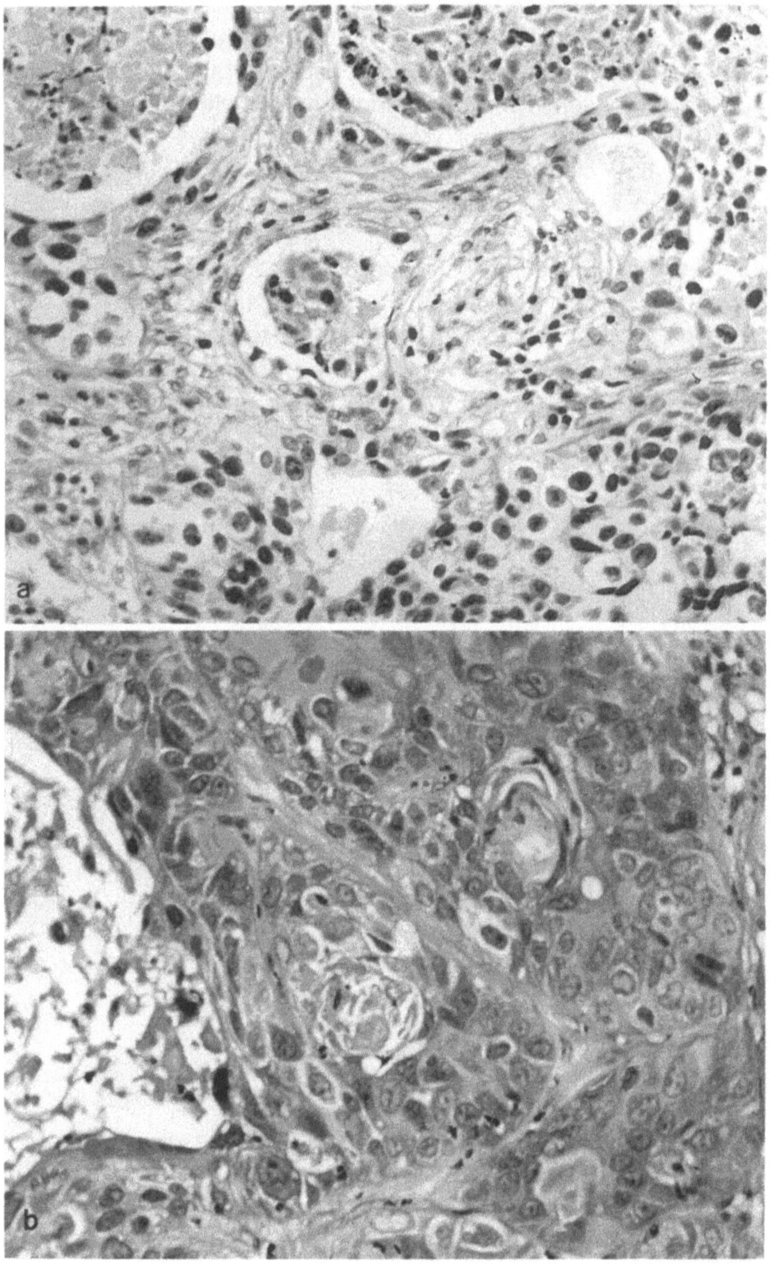

Fig. 2. **a** CX143 adenosquamous tumour showing moderately differentiated squamous elements in original passage. **b** CX143 6th passage showing poorly differentiated adenocarcinomatous elements

Table 2. Maximum tolerated doses of drugs used in chemotherapy studies

Drug	Dose (mg/kg)	Schedule
Vindesine (VDS)	2	d 1
Cyclophosphamide (CTX)	150	d 1
Ifosfamide (IFOS)	300	d 1
etoposid (VP16)	15	d 1-3
Cisplatin (DDP)	7	d 1
Carboplatin (CBDCA, JM8)	75	d 1
Ethylenediamineplatinum II malonato (JM40)	40	d 1

Discussion

This study describes the establishment of human non-small cell lung cancer xenografts in immune-deprived mice. Xenograft lines were successfully established in 15/44 tumours (34%), this figure being much lower than described by Shorthouse (5). There are two likely reasons for this: firstly the majority of samples in this study were obtained from the primary tumours, whereas metastatic deposits were commonly used in the study reported by Shorthouse, with the success rate from metastases known to be higher. In contrast, the take rates in subsequent xenograft passages were similar to

Table 3. Chemotherapy studies in human non-small cell lung cancer xenografts

Xenograft	Histology	Doubling time (days)	VDS	CTX	VP16 (SGD)	IFOS	DDP	JM8	JM40
NX002	MD[a] squam	16.5	<0.2	0.36	0.2	NT[b]	2.0	0.9	0.33
CX109	PD squam	18.6	0.31	NT	NT	NT	0.73	0.73	0.36
CX117	PD adeno	11.25	<0.2	0.52	0.91	1.4	0.87	0.95	0.76
CX133	MD squam	14.0	0.26	0.41	<0.2	0.47	1.23	1.16	0.71
CX140	MD squam	12.6	<0.2	NT	<0.2	0.58	1.45	NT	1.33
CX143	adenosquam	19.1	0.21	0.63	0.38	1.38	0.29	0.48	0.48

[a] MD = moderately differentiated PD = poorly differentiated.
[b] NT = not tested.

those described by Shorthouse, indicating that any immunological differences between the different populations of mice used were unlikely to have been a factor in the establishment of tumour lines.

Histologically, these tumours maintained the characteristics of the primary tumour, although minor changes in differentiation were observed. Both increased and decreased differentiation were seen with increased differentiation more commonly observed than previously described (11). The chemotherapy studies in these xenografts showed minimal single agent activity mirroring clinical experience in non-small cell lung cancer. Of the single agent activity, cisplatin appeared to be the most active with less activity noted with 2 platinum analogues.

Human lung tumour xenografts therefore offer excellent models for the study of the biology and therapeutic responsiveness of these tumours. Squamous carcinoma xenografts are of particular interest in view of the relative scarcity of squamous lung cancer cell lines. There are limitations to the use of these xenografts, since there is variability in growth rate even within 1 xenograft passage, and spontaneous regressions of tumour occur in 5-10% of mice, although these are generally observed in smaller tumours. Spontaneous regression is less of a problem; however, there are advantages to the use of artificially immune-deprived mice, in that SPF conditions are not required. Despite this, lung xenografts using nude or artificially immune-deprived mice offer an excellent model for the screening of new agents, although for detailed assessment of chemosensitivity large numbers of mice may be required.

Acknowledgements

We would like to thank the physicians and surgeons involved in obtaining the tumour samples including Dr. I.W.B. Grant and Dr. G.K. Crompton, Respiratory Diseases Unit, Northern General Hospital, Edinburgh and Mr. P.R. Walbaum, Mr. E.Cameron and Mr. Sang, Thoracic Surgery Unit, City Hospital, Edinburgh, and especially Drs. Lamb and McIntyre of the University and Western General Hospital Departments of Pathology for their help with these studies.

References

1. Gralla RJ, Casper ES, Kelsen DP et al: Cisplatin and vindesine combination chemotherapy for advanced carcinoma of the lung: A randomised trial investigating two dosage schedules. Ann Intern Med 1981 (95): 414-420
2. Woods RL, Levi JA, Page J et al: Non-small cell lung cancer: A randomised comparison of chemotherapy with no chemotherapy. Proc Am Soc Clin Oncol 1985 (4):C-691
3. Carmichael J, Gregor A, Cornbleet MA et al: Vindesine and cis-platinum in the treatment of non-small cell lung cancer. Eur J Cancer Clin Oncol 1985 (21): 811-814
4. Steel GG: The growth and therapeutic response of human tumours in immune deficient mice. Bull Cancer 1978 (65): 465-472
5. Shorthouse AJ, Smyth JF, Steel GG et al: The human tumour xenograft - a valid model in experimental chemotherapy. Br J Surg 1980 (67): 715-722
6. Mattern J et al: Different growth rates of lung tumours in man and their xenografts in nude mice. Eur J Cancer Clin Oncol 1980 (16): 289-291
7. Steel GG, Courtenay VD, Rostom AY: Improved immune-suppression techniques for the xenografting of human tumours. Br J Cancer 1978 (37): 224-230
8. Kopper L and Steel GG: The therapeutic response of three human tumor lines maintained in immune-suppressed mice. Cancer Res 1975 (35): 2704-2713
9. WHO: International histological classification of tumours. No 1. In: Histological typing of lung tumours. Geneva: World Health Organisation 1981 2nd edition
10. Lamb D: The pathology and classification of lung cancer. In: Smyth JF (ed) The management of lung cancer. London: Edward Arnold 1984: 19-35
11. Shorthouse AJ. MS Thesis, University of London 1981

Establishment and Response to Chemotherapy of Human Cervical Carcinoma Xenografts

L. R. KELLAND and K. S. TONKIN

Radiotherapy Research Unit, Institute of Cancer Research, Clifton Avenue, Sutton, Surrey SM2 5PX, United Kingdom

Abstract

Nine serially transplantable xenograft lines of human cervical carcinoma have been established in nude mice from 23 original biopsy specimens. The lines represent the spectrum of patient ages (30 to 80 years) for this disease. Tumours of poor, moderate and well differentiated histologies are represented; these morphological characteristics were well maintained from patient biopsy through to xenograft. Colony forming efficiencies assessed using the Courtenay soft agar assay were very low for all of the lines with a maximum of only 0.04%. Considerably higher efficiencies were observed using an anchorage dependent monolayer assay (maximum 30%). The *in vivo* response of 3 of the lines to the chemotherapeutic agents cisplatin (5 and 8 mg/kg), etoposide (45, 60 and 90 mg/kg) and bleomycin (150 mg/kg) showed them to be generally unresponsive with the maximum specific growth delay being only 1.48 using bleomycin for line HX155. However, there was evidence for individuality of chemosensitivity to these agents within this tumour type.

Introduction

Carcinoma of the uterine cervix is the sixth most common cause of cancer death in women today, accounting for up to 1% of all female deaths in the Western world (7). Although the primary treatment consists of radiotherapy and surgery there is increasing awareness of the need for active chemotherapeutic agents to be found for the more advanced stages of this disease (19). In order to investigate the single agent chemosensitivity of this disease in the laboratory we have attempted to establish and use human tumour xenografts transplanted in nude mice, in the light of experimental data supporting the view that xenografts broadly maintain the level of chemotherapeutic responsiveness of the source tumours in man (17). We here describe the successful establishment of 9 serially transplantable xenograft lines and the *in vivo* response of 3 lines to etoposide, bleomycin and cisplatin.

Materials and Methods

Establishment of Xenografts from Tumour Biopsies

Between November 1984 and May 1985, 23 biopsy specimens were collected from patients presenting at the Radiotherapy Department of the Royal Marsden Hospital for initial staging and treatment planning of their disease. Biopsies were held for 2 hours in ice-cold Ham's F12 medium containing penicillin (10^5 units/l), streptomycin (100mg/l) and neomycin (10mg/l). Specimens were finely chopped with crossed scalpels and rinsed in phosphate-buffered saline (PBS). Pieces of 2 mm^2 were then implanted subcutaneously into the flanks of nude mice. Female nude mice (Nu/Nu) were used for these studies and housed in plastic film isolators. One day prior to tumour implantation the mice received 4 Gy total body irradiation from a ^{60}Co source. When the xenografts had reached a diameter of 10-15 mm they were excised, chopped and transplanted in an attempt to establish a tumour line.

Determination of Chemosensitivity

Xenografts from passage 4 to 8 maintained in female nude mice (nu/nu) were used. Caliper measurements of diameters were carried out at least every 5 days until tumour weights (calculated using a calibration curve technique, see ref. 17) exceeded 4 times that of control tumours.

Tumours that reached a diameter of 6-9 mm were selected and randomized into control and test groups each containing at least 5 tumours. Drugs were administered by intraperitoneal injection at their maximum tolerated dose (LD_{10}) as determined by preliminary experiments. Control mice received injections of PBS. Specific tumour growth delay (SGD) was used as the endpoint of all chemotherapy studies:

$$SGD = \frac{T_2 - T_1,}{T_1}$$

where T_1 is the time in days for control tumours to double in volume and T_2 is the time in days for treated tumours to double in volume.

Soft Agar Cloning from Xenografts

Following cervical dislocation of mice, tumours of 0.2–0.8 g in weight were excised. They were finely chopped, rinsed in PBS and disaggregated at 37 °C C for 30 min in Ham's F12 medium containing collagenase 0.2 mg/ml, pronase 0.5 mg/ml and deoxyribonuclease I 0.2 mg/ml. Single-cell suspensions were prepared by centrifuging (100 g, 5 min), resuspending in medium containing 15% bovine calf serum and filtering through a 20 µm polyester mesh. Cloning efficiencies were assessed according to the method of Courtenay & Mills (2) using August rat erythrocytes and heavily irradiated (HR) tumour cells in 0.3% agar.

Results

From the original 23 biopsy specimens, 9 gave rise to serially transplantable xenograft lines representing a success rate of 39%. The patient details for the 9 xenograft lines are described in Table 1. Assessment of tumour doubling times at passage 4 revealed a wide variation ranging from about 6 days for HX151 and HX155 to about 23 days for HX160 and HX157. Generally the tumours appeared to be well vascularized and often when their average weight exceeded 0.8 g, cystic or necrotic areas were apparent. The cystic nature and the slow tumour growth of many of the xenografts limited the chemosensitivity experiments to 3 lines; HX155, HX156 and HX160. As can be seen from Table 1 these lines not only represent the spectrum of patient ages for this disease but also represent poorly, moderately and well differentiated tumour histologies.

The maintenance of biological characteristics in the xenografts was assessed by comparing the histology of xenografted tumour sections with sections of the original patient biopsies. Representative haematoxylin and eosin sections for HX155, HX156 and HX160 are shown in Figure 1 (a–f). The Figure shows that the morphological appearance of the tumours as xenografts in nude mice was generally indistinguishable from that of the original patient biopsy. HX155 and HX160 maintain the morphology of well differentiated keratinizing tumours in nude mice whereas HX156 appears as a poorly differentiated tumour.

Table 1. Summary of patient details for human cervix carcinoma xenografts

Line designation	Clinical stage of disease[a]	Histology	Age of patient (years)	Tumour doubling time at passage 4 (days)
HX151[c]	IB	Poorly differentiated SCC[b]	30	6
HX155[c]	IB	Moderately differentiated infiltrating keratinizing SCC	44	7
HX156[c]	IIB	Poorly differentiated SCC	31	18
HX157	IIB	Well differentiated keratinising SCC	78	23
HX158	IIIB	Poorly differentiated SCC	47	22
HX159	IB	Well differentiated SCC	33	22
HX160[c]	IIB	Moderately well differentiated SCC	63	23
HX162	IIB	Poorly differentiated SCC	31	14
HX166	Recurrence from 12 years previous	Poorly differentiated	80	12

[a] Clinical staging according to FIGO (Cancer Committee of the International Federation of Gynecology and Obstetrics).
[b] SCC, Squamous cell carcinoma.
[c] Also established as an in vitro continuous cell line.

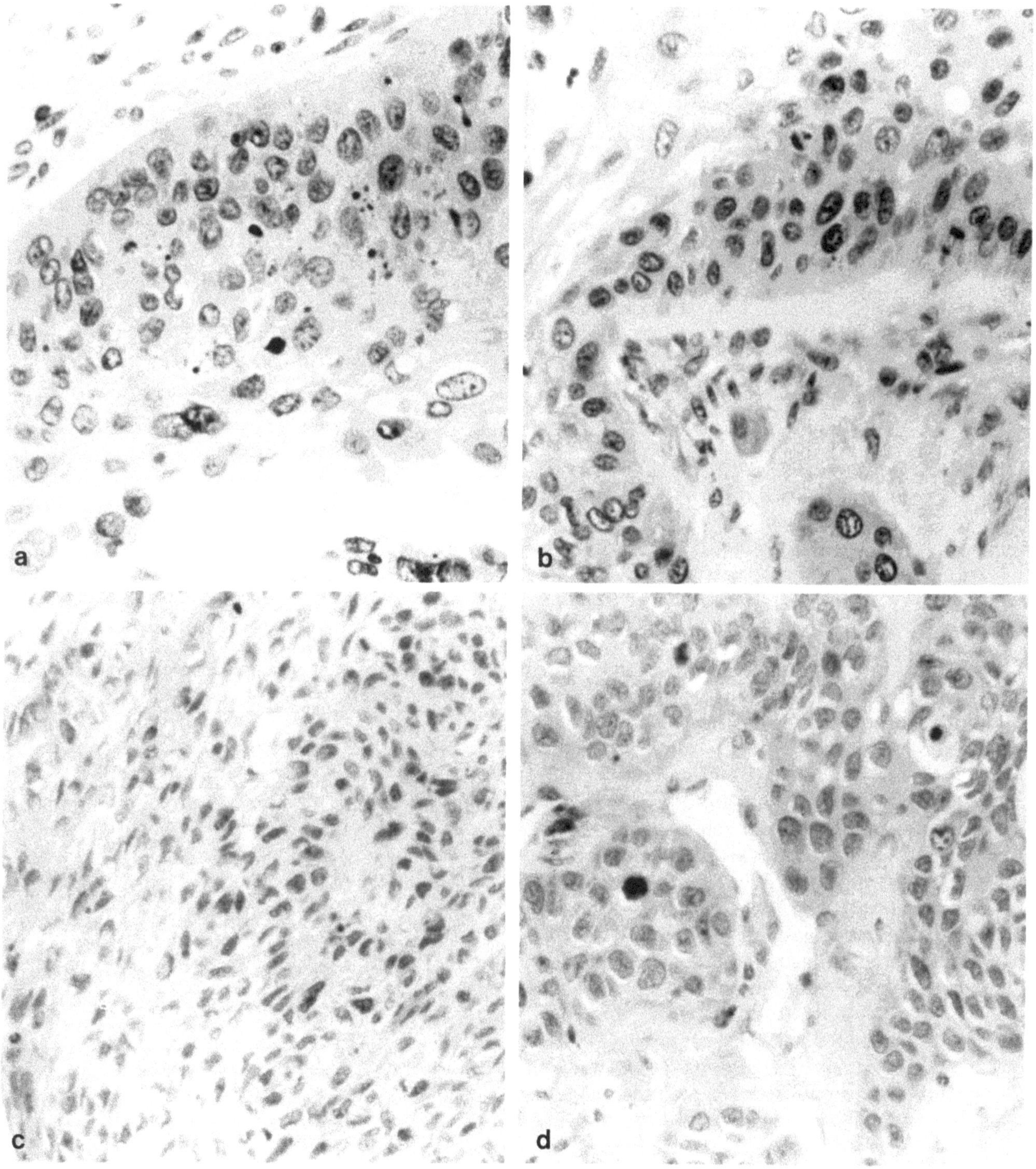

Fig. 1 a–f. Histology of sections of tumours from either original cervical carcinoma patient biopsies (**a, c, e**) or from xenografts in nude mice (passage 4) (**b, d, f**); HX155 (**a** and **b**), HX156 (**c** and **d**), HX160 (**e** and **f**). H & E ×250

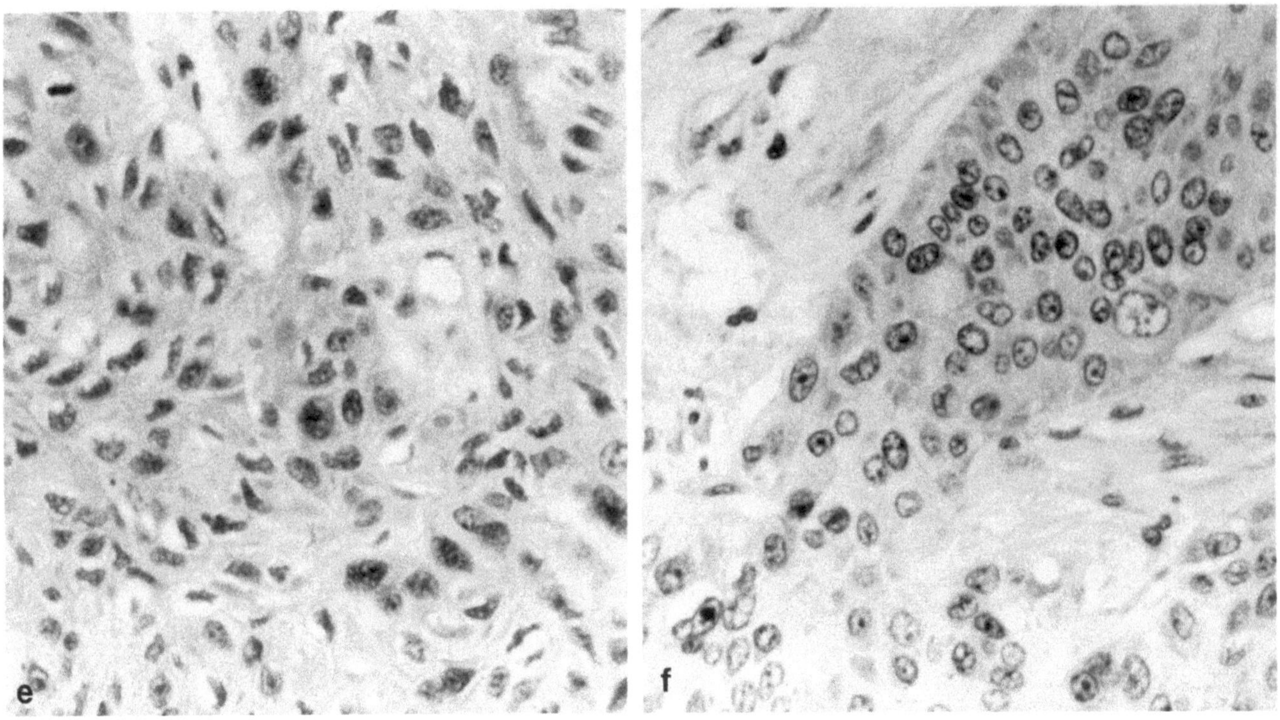

Fig. 1 e–f.

Colony-forming efficiencies were assessed in soft agar according to the method of Courtenay & Mills (2) for the 9 xenograft lines. In all cases the efficiencies were very low with HX151, HX156 and HX160 giving values of around only 0.04% and the remainder producing values of less than 10^{-3}.

In addition to the established xenograft lines, 4 of the original biopsies produced independent continuous cell lines, the details of which have been published separately (10). For these lines a monolayer cloning assay was developed involving the use of 3T3 mouse fibroblast feeder layer support. This anchorage-dependent assay produced cloning efficiencies that were much higher than those described herein for the soft agar assay with HX156 giving 30%, HX151, HX160, 12% and HX155 20%.

In our initial experiments we have used the chemotherapeutic agents which have proved most useful in clinically advanced stages of the disease (6). A summary of the results we have obtained using etoposide (45, 60 and the maximum tolerated dose of 90 mg/kg), bleomycin (150 mg/kg) and cisplatin (5 and 8 mg/kg) is shown in Table 2. In the case of etoposide we have compared the effect of a single administration versus 3 doses given daily over 3 days. Results show that, for all 3 lines, the response was poor with the repeated doses as well as with the single dose. Although the responses with etoposide were poor there was evidence of individuality within this tumour type with HX155 being consistently more chemosensitive than HX156 or HX160.

Table 2. Summary of chemotherapeutic response for carcinoma of the cervix xenografts

Drug	Dose[a] (mg/kg)	Specific tumour growth delay[b]		
		HX155	HX156	HX160
Etoposide	45	0.6	-	0.17
	15 × 3	1.1	-	0.25
	60	1.18	0.58	0.50
	20 × 3	0.70	0.25	0.12
	90	-	0.68	-
	30 × 3	-	0.43	-
Bleomycin	150	1.48	0.23	-
Cisplatin	8	0.93	-	-
	5	0.53	-	-

[a] Multiple doses were given at daily intervals.

[b] Specific tumour growth delay $= \frac{T2 - T1}{T1}$ as described in materials and methods.

The results with bleomycin showed that the xenografts were again generally unresponsive. However, HX155 was again more responsive than HX156. Experiments using HX155 with cisplatin at the maximum tolerated dose of 8 mg/kg also resulted in little specific growth delay although this line was the most sensitive to the effects of etoposide.

Discussion

We have achieved a success rate of 39% in establishing serially transplantable xenografts of human squamous cell carcinomas of the cervix. From Table 1 and from an analysis of the staging and histologies of all 23 biopsies studied, there was no obvious correlation between growth as a human tumour line and stage of clinical disease or the degree of differentiation of the tumour. In addition to the 9 human tumour lines, 4 tumours were also independently established as continuous cell lines, the biological details of which have been published separately (10).

There have been few previous reports of the establishment of serially transplantable xenografts of human cervix carcinoma in immune-deficient mice. Krishan et al. (12), investigating the growth of gynecological tumours in nude mice, achieved a success rate of 33% (4 of 12) for cervix carcinoma which included both squamous and adenocarcinomas. In another study (13) a success rate of 11% (1 of 9) was observed using unirradiated nude mice, whereas 25% (3 of 12) showed progressive growth in mice given 4 Gy gamma-irradiation. We have also observed an increase in take-rates in nude mice given 4 Gy total body irradiation prior to implantation (unpublished data). In addition to these 2 studies there have been some carcinoma of the cervix xenografts established individually and not part of an overall investigation into the growth of this disease (1, 11, 20) including a line established from Hela cells (18).

Within this Department we have achieved much success in the cloning of human tumour cells using the Courtenay soft agar assay (2), including melanomas (9), pancreatic carcinomas (9), neuroblastomas (4), ovarian adenocarcinomas (8) and lung carcinomas (5). Our results using this assay with the cervix carcinoma xenografts have been disappointing with none of our lines producing a colony forming efficiency of greater than 0.04%. However, these results are perhaps not surprising in the light of investigations into the soft agar cloning ability of other squamous cell carcinoma cell lines (notably epidermis, tongue and head and neck) which have also shown very low colony forming efficiencies (3, 14, 15).

Our results on the *in vivo* response to chemotherapy of the 3 xenograft lines show the cervix carcinomas to be generally unresponsive. The maximum specific growth delay (SGD) (volume doubling times saved by a single drug dose) observed was only 1.48 using bleomycin in HX155. However, it is possible that the SGD in line HX160 using bleomycin might be greater. The unresponsiveness of the xenografts is in keeping with the poor clinical responses reported when chemotherapy is used in advanced disease (e.g., 6, 19). To date there is not enough clinical data on the primary treatment of carcinoma of the cervix with chemotherapy to be confident of the responsiveness. However, our data do lend support to the view that differences in chemosensitivity exist between individual tumours as, in the case of etoposide (60 mg/kg), there was a clear difference in response between HX155 (SGD of 1.18) and HX156 and HX160 (SGD values of 0.58 and 0.50, respectively). Evidence for such individuality within tumour types has been observed previously in colo-rectal carcinoma, breast carcinoma and bronchial carcinomas (17). Using etoposide administered as a single dose or given as 3 daily doses show that, at doses of 60 and 90 mg/kg, growth delay was slightly greater for the single dose. For the lower dose of 45 mg/kg there was a trend to indicate that the opposite was the case. However, these results comparing single versus 3 daily doses are not significantly different.

The general unresponsiveness of the cervical carcinoma xenografts to 3 drugs thought to be among the most promising clinically (6) may be due to a number of reasons: cervix carcinoma cells could be inherently resistant to chemotherapeutic agents presently available, or the uptake of these agents into xenografts may be poor. We are currently investigating the *in vitro* chemosensitivity of 4 carcinoma of the cervix cell lines to these and other agents. The cervical carcinoma xenografts described in this study represent a panel of human tumour lines available for the laboratory evaluation of new chemotherapeutic agents for the treatment of this disease.

Acknowledgements

Our thanks are due to Dr. Joan Baker and the Radiotherapy staff for the supply of surgical biopsies. Thanks are also due to Dr. Gordon Steel for helpful discussion, Professor Barry Gusterson for histopathology appraisals, Mr. Ted Merryweather for care of the animals, Miss P. A. Wilson and Miss L. Burgess for technical assistance and Mrs. S. Stockbridge and Miss R. Couch for typing the manuscript. This investigation was supported by NCI grant R01 26059.

References

1. Benjamin I, Rana MW and Freel J: Radiosensitizing effect of misonidazole on human squamous cell carcinoma of cervix transplanted in nude mice. In: 'Immune-Deficient Animals' 4th Int Workshop on immune-deficient animals, Exp Res 1984 Karger, Basel pp 389-394

2. Courtenay VD and Mills J: An in vitro colony assay for human tumours grown in immune-suppressed mice and treated in vivo with cytotoxic agents. Br J Cancer 1978 (37): 261-268
3. Cowley G, Gusterson B and Knight J: Growth in agar and tumor formation in immunologically incompetent mice as criteria for keratinocyte transformation. Cancer Lett 1983 (21): 95-104
4. Deacon JM, Wilson PA and Peckham MJ: The radiobiology of human neuroblastoma. Radiother Oncol 1985 (3): 201-209
5. Duchesne G, Eady J, Peacock J and Pera M: A panel of human lung carcinoma lines: Establishment, properties and common characteristics. Br J Cancer 1987 (56): 287-293
6. Friedlander ML, Atkinson K, Coppleson JVM, Elliot P, Green D, Houghton R, Solomon HJ, Russell P and Tattersall MHN: The integration of chemotherapy into the management of locally advanced cervical cancer: A pilot study. Gynecol Oncol 1984 (19): 1-7
7. Hill GB: 1975 World Health Statistics Report (28): 323
8. Jones AC, Stratford IJ, Wilson PA and Peckham MJ: In vitro cytotoxic drug selectivity testing of human tumour xenografts grown as multicellular tumour spheroids. Br J Cancer 1982 (46): 870-879
9. Kelland LR and Steel GG: Dose-rate effects in the radiation response of four human tumour xenografts. Radiother Oncol 1986 (7): 259-268
10. Kelland LR, Burgess L and Steel GG: Characterization of four new cell lines derived from human squamous carcinomas of the uterine cervix. Cancer Res 1987 (47): 4947-4952
11. Kleine W, Wrzodek W, Stange S and Ladner HA: Hochvoltbestrahlung xenotransplantierter menschlicher Ovarial-endometrium- und Zervixkarzinome. Strahlentherapie 1981 (76): 289-295
12. Krishan A, Haines H and Averette HE: Gynecological oncology studies in nude mice. In "The nude mouse in experimental and clinical research" Vol 2, Fogh and Jorgen 1982: 451-459 Academic Press, New York
13. Maruyama Y, Feola JM and Nagell JR: Heterotransplantation of human cervix cancers in radiation-conditioned or non-conditioned athymic mice. Cancer 1985 (56): 2635-2639
14. Rheinwald JG and Beckett MA: Tumorigenic keratinocyte lines requiring anchorage dependence and fibroblast support cultured from human squamous cell carcinomas. Cancer Res 1981 (41): 1657-1663
15. Rupniak HT and Hill BT: The poor cloning ability in agar of human tumour cells from biopsies of primary tumours. Cell Biol Int Rep 1980 (4): 479-486
16. Steel GG: In "The Growth Kinetics of Tumours" 1977. Oxford University Press
17. Steel GG, Courtenay VD and Peckham MJ: The response to chemotherapy of a variety of human tumour xenografts (Review). Br J Cancer 1983 (47): 1-13
18. Tokita H, Tanaka N, Sekimoto K, Ueno T, Okamoto K and Fujimura S: Experimental model for combination chemotherapy with metronidazole using human uterine cervical carcinomas transplanted into nude mice. Cancer Res 1980 (40): 4287-4294
19. Ward BG, Shepherd JH and Monaghan JM: Occult advanced cervical cancer. Br Med J 1985 (290): 1301-1302
20. Xynos FP, Benjamin I, Sapiente R, Rana MW and Nalesnik WJ: Adriamycin and hydroxyurea as radiopotentiators in the treatment of squamous cell carcinoma of the cervix implanted in nude mice: A preliminary report. Gynecol Oncol 1980 (9): 170-176

The Usefulness of the Nude Mouse Model for Predicting the Chemosensitivity of Human Tumors. Studies in Malignant Melanoma

Ø. FODSTAD, S. AAMDAL, and A. PIHL

Institute for Cancer Research, The Norwegian Radium Hospital, Montebello, 0310 Oslo 3, Norway

Abstract

The usefulness of human tumor xenografts in experimental chemotherapy depends on the representativity of the xenografts used for the respective types of clinical cancer. The overall experience in malignant melanoma has provided strong evidence in support of the validity of the nude mouse xenograft system for chemosensitivity testing. A sensitivity profile obtained in a panel of 5 melanoma xenografts seemed to reflect the clinical situation, and a pattern of tumor response to CCNU and vinblastine similar to that seen in the clinic was reproduced in xenografts. Moreover, a clear relationship between tumor size and response to vinblastine was demonstrated. In screening new drugs, short-term assays may be useful as prescreens, but the need for long-term experiments in nude mice is emphasized. A new drug, Mitozolomide, was selected in preclinical Phase II studies in xenografts for evaluation in melanoma patients, and since clinical activity of this drug was later established, the results support a more extended use of this approach. Furthermore, it is proposed that analogs of established drugs should undergo careful preclinical selection before being introduced in the clinic.

Introduction

Human malignant melanoma is one of the cancer types most extensively studied as xenografts in nude mice (1-6). It is known that xenografted tumors by and large retain the most important biological characteristics of their parent tumors (2, 4, 6, 7), and there is strong evidence suggesting that the chemosensitivity of human melanomas is reflected in melanoma xenografts. Firstly, individual tumor lines have been found to respond to drugs which are clinically active in malignant melanoma (1, 2). Secondly, when panels of xenografted tumors were tested for sensitivity to various drugs, the chemosensitivity profiles were found to be similar to those seen for melanomas in the clinical situation (5, 6). Thirdly, direct comparisons have been made between drug effects against xenografts and the response to the same treatment in the patients from whom the xenografts originated (6, 8, 9). Meaningful comparisons of this type are hard to make. This is due to the problems involved in establishing xenograft lines and to possible difficulties in correlating the assessment of response in the two systems. In addition, it is clinically only possible to test one drug regimen before the tumor may change, acquiring characteristics different from those of the xenograft. Lastly, when a new compound with promising effects in xenografts (10) is later found to be clinically active (11), this is also an indication that the chemosensitivity of human melanomas is reflected in xenografts.

In this paper, examples of data supporting the validity of the melanoma xenograft system are presented, and areas of use of the model in experimental chemotherapy are discussed.

Activity of Established Drugs

Chemosensitivity Profiles

For the purpose of testing the chemosensitivity of a particular tumor type, the inter-tumor heterogeneity makes it absolutely necessary to study panels of xenografts. In malignant melanomas such panels should cover a range of characteristics with respect to growth rates, melanin production and sensitivity to drugs, in order to obtain data that adequately reflect the sensitivity of clinical melanomas. Such studies have been performed by several investigators (5, 6, 8, 9) and the results obtained are in agreement with the clinical experience in melanoma.

In Table 1 we have listed the growth rates of 5 tumor lines together with their response to 4 cytostatic drugs. The tumor volume doubling times of the

Table 1. Effect of chemotherapy on human melanoma xenografts. Groups of 8-10 animals were treated once weekly x 3 with DTIC (250 mg/kg i.p.), CCNU (20 mg/kg i.p.), vinblastine (2.5 mg/kg i.v.), or ricin (1.5 µg/kg i.v.).

Xenograft	Tumor volume doubling time (days)[a]	Specific growth delay[b]			
		DTIC	CCNU	Vinblastine	Ricin
GEX	9.5 +/− 0.7	2.6	2.7	0.1	0.5
MFX	8.9 +/− 0.7	0.1	0.3	1.5	0.3
EFX	6.3 +/− 0.7	1.0	1.1	1.1	0.6
VNX	3.3 +/− 0.1	0.3	0	0.3	0
EEX	3.0 +/− 0.3	0.5	0.1	1.4	1.5

[a] Tumor volume doubling time (30-60 mm^3), TD, calculated as described in ref. 46 on the basis of at least 3 experiments, each involving 8-10 mice.

[b] Specific growth delay $= \frac{TD_T - TD_C}{TD_C}$; (see 20). TD_T=TD in the treated group, TD_C=TD in the control group.

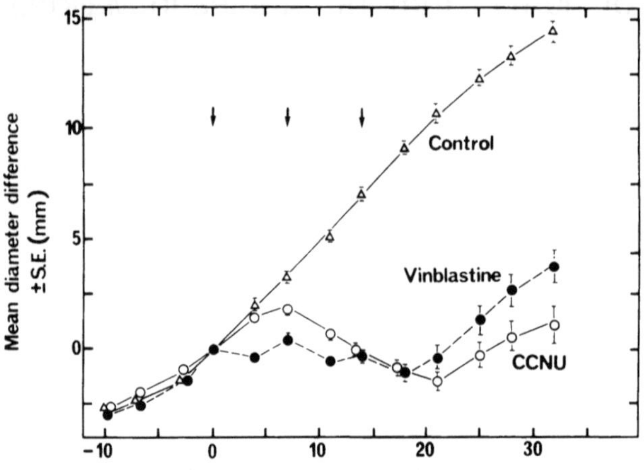

Fig. 1. Effect of vinblastine and CCNU against the THX human melanoma line in nude mice. Groups of 10-16 mice were treated at the times indicated by the arrows either with 2.5 mg/kg of vinblastine i.v., 20 mg/kg of CCNU i.p., or they were used as untreated controls. For each tumor 2 diameters were measured twice weekly, and the mean diameter difference for each group was calculated as described previously (6, 19)

different lines varied from 3.0 to 9.5 days. As expected, the effect of each drug differed from tumor to tumor, and each tumor showed dissimilar sensitivities to different drugs. Two of the xenografts, MFX and VNX, were obtained from patients previously treated with DTIC and CCNU until they developed resistance, and both xenografts were unresponsive to these drugs (6). VNX showed cross-resistance to vinblastine and ricin, whereas MFX had retained some sensitivity to vinblastine. It is not likely that this panel of only 5 tumors is fully representative of melanomas in general (12), but the data in Table 1 illustrate how the representativity of such xenograft panels can be examined in regard to sensitivity to drugs.

Patterns of Tumor Response

Differences between anti-cancer agents in action mechanisms and pharmacokinetics lead to differences in lag times before effects caused by the drug are observed, as well as in the duration of these effects. Thus, it is known that nitrosoureas, e.g. CCNU, have a delayed type of activity and bone marrow toxicity in the clinic, and hence must be administered at longer intervals than those common for most other drugs. Vinblastine, on the other hand, becomes effective more rapidly and may be given in more frequent schedules. We have studied in nude mice the antitumor effects of CCNU and vinblastine, given once weekly for 3 weeks, against THX melanoma xenografts. In repeated experiments the pattern of response was identical. Data from one such experiment are shown in Fig. 1. It can be seen that the effect of CCNU became first apparent after the second course of treatment, after which the tumor decreased in size until about one week post-treatment, and then growth resumed. In the case of vinblastine, the first treatment course induced a rapid decrease in tumor size, apparent already 4 days after the administration of the drug. However, 3 days later the tumour size had increased again, almost to pre-treatment level. The subsequent 2 treatments produced a similar pattern of effect, but the overall tumor volume reduction was nevertheless about the same as for CCNU, before the tumors started to grow again at approximately the same rate as the untreated controls. In conclusion, the characteristic delayed and rapid response to CCNU and vinblastine, respectively, as seen in the clinic can be reproduced in xenografts, and supports the validity of testing drugs in the nude mouse system.

Relationship Between Tumor Size and Response

It is a well known clinical experience that small tumors often respond better to therapy than larger tumors. This may be explained by the lower growth fraction in large tumors (13), as well a by differences in blood supply and drug access to the malignant cells between tumors of different size. When the data for the Vinblastine-treated mice in Fig. 1 were split up into 3 different subgroups according to the size of the xenografts at the start of treatment, an interesting pattern was seen (Fig. 2). First-

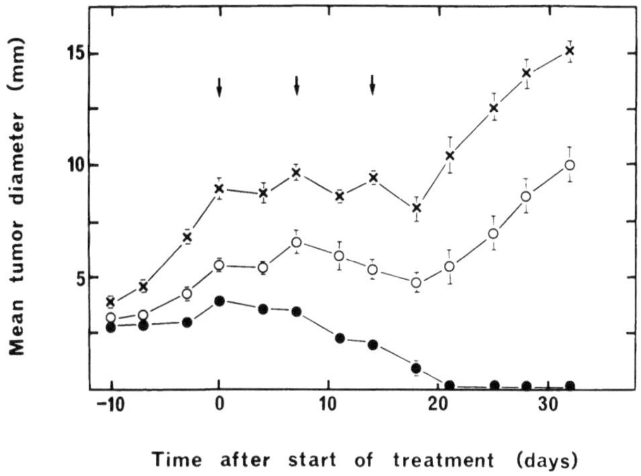

Fig. 2. Relationship between tumor size and response to vinblastine of the THX melanoma line. The xenografts in the 16 Vinblastine-treated mice in Fig. 1 were divided into 3 groups according to size at start of treatment, with mean tumor diameters of 9.0 mm (5 tumors) (×) 5.5 mm (8 tumors) (○) and 3.9 mm (3 tumors) (●), respectively. Treatment and tumor measurements as described in Fig. 1

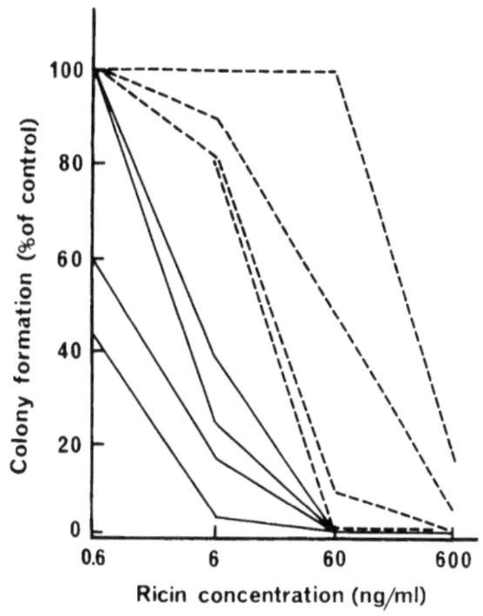

Fig. 3. Dose-response curves *in vitro* of tumor cells prepared from 4 human adenocarcinomas of the lung (———) and of 4 human melanomas (------), treated with increasing concentrations of ricin. Single-cell suspensions prepared from xenografts were exposed for 1 hour at 37 °C, and the residual ability to form colonies in soft agar was measured

ly, 5 relatively large xenografts with a mean diameter of 9 mm when treated with the first dose of vinblastine, responded as described above. A subgroup of 8 tumors with a mean diameter of 5.5 mm showed a similar response, except that between the second and third treatment course there was no indication of regrowth of the tumors (Fig. 2). In the last subgroup of 3 mice with very small but clearly growing xenografts (mean diameter of 3.9 mm), no regrowth was seen between treatments. Moreover, the tumors continued to shrink after the end of therapy until they became unmeasurable, and growth did not resume during an observation period of 45 days. The results demonstrate a clear relationship between tumor size and response, similar to the experience in the clinic. Furthermore, the data suggest that the reasons for the observed differences may be related to different abilities of malignant cells within tumors of different sizes to recover from the damage caused by the drug. It is conceivable that a larger fraction of G0/G1 cells present in large tumors may avoid damage and repopulate the tumor in the interval between 2 drug doses.

Activity of New Drugs

The Role of Nude Mice Models in Screening

The validity of chemotherapy experiments in the nude mouse model documented for established anticancer agents would suggest the use of this system also for screening the efficacy of new compounds. However, the use of s.c. xenografts for screening purposes is limited by the fact that such experiments are expensive and very time-consuming. Therefore, large-scale screening must be based on short-term *in vitro* and *in vivo* prescreens, in order to select a reasonable number of promising drugs for testing in the nude mouse model (10, 14). However, one should bear in mind that there are limitations with respect to pharmacology and toxicity inherent in any *in vitro* assay. The need for *in vivo* confirmation of *in vitro* positive results of testing of new drugs is shown by our experience with ricin, an investigational drug which acts by inhibiting protein synthesis (15). Following a Phase I clinical trial (16), we obtained *in vitro* data suggesting that this compound could be active against adenocarcinomas of the lung. In Fig. 3 dose-response curves in the Human Tumor Colony Forming Assay (HTCFA) for 4 human melanomas and 4 adenocarcinomas are shown. The latter tumors were all much more sensitive to ricin than the melanomas, resulting in a difference in mean ID_{50} values for these tumor types of approximately 2 logs. Ricin was inactive in 3 melanoma patients (16) but had previously shown moderate antitumor effects in some melanoma xenografts, and on the basis of the HTCFA results it seemed reasonable to expect adenocarcinomas of the lung to be much more sensitive *in vivo* than melanomas. However, no *in vivo* response to ricin was seen in such tumors, neither in a Phase II clinical study nor in nude mice (unpublished). These findings emphasize the problems of interpreting the *in vivo* relevance of data obtained *in vitro*.

Comparison of Drug Analogs

Analog development and testing have attracted a lot of attention during the last 10 years, partly due to the lack of new leads in the search for better anticancer drugs. It is usually difficult to evaluate whether an analog has an improved therapeutic index compared to the parent compound. In general such assessments have been performed in clinical trials rather than in proper preclinical models. Considering all the inherent uncertainties of such clinical studies, often performed as multi-center trials and in heterogeneous patient populations, it takes time to obtain definite evidence whether or not the analog is better than the parent drug. Thus, it seems reasonable to assume that a significant part of the resources spent on evaluation of analogs could have been saved if such comparisons had been performed preclinically in nude mouse systems (17). To illustrate this, we have compared the effect against 2 melanoma tumor lines of equitoxic doses of Mitozolomide, a new clinically active imidazotetrazine (10, 11, 18), with that of its analog Metazolomide. The results in Fig. 4 show that both drugs were equally active against the GEX tumor (left panel), resulting in a decrease in tumor size. In the THX II tumor (right panel), Metazolomide had no effect, whereas Mitozolomide definitely retarded the growth of the tumor. If the present data is also valid in other tumors, and there are no differences in drug formulation or in toxicity in favor of Metazolomide, there seems to be no good reason to develop this analog any further. The data in Fig. 4 show that Mitozolomide is clearly more active than CCNU. This is an interesting feature as there are indications of cross-resistance between these drugs (10).

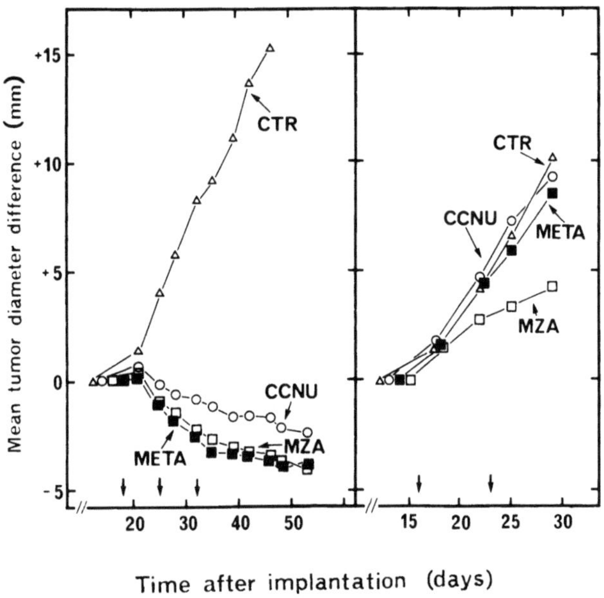

Fig. 4. Effect of 3 chemotherapeutic drugs against the GEX (left panel) and THX II (right panel) human melanomas in nude mice. Groups of 8–10 mice were treated at the times indicated by the arrows with 20 mg/kg of CCNU i.p., 30 mg/kg of Mitozolomide i.v., 175 mg/kg of Metazolomide i.v., or used as untreated controls. Measurements and calculations as in Fig. 1

Preclinical Phase II Studies

The selection of drugs for introduction into clinical trials was previously based on anti-tumor data obtained with murine tumor models. These systems have not given any indications of tumor-type specific efficacy against human cancer. As a result, clinical Phase II studies have been performed rather randomly, in the hope of obtaining signs of activity in some tumor types. This has led to the frustrating experience that most of these trials were completely negative in solid tumors. The usefulness of xenografts for *pre*clinical Phase II trials is discussed elsewhere in this Monograph (see chapter 9). Here we will just mention one successful example on the use of xenografts in this respect. As previously reported (10), Mitozolomide proved to have promising effects against human melanoma xenografts in nude mice. With that as a basis we initiated a clinical Phase II study with Mitozolomide in this tumor type. The results showed that the drug is in fact clinically active against melanoma (11). Together these findings support the validity of this approach in selecting drugs for clinical trials.

Acknowledgements

The authors wish to thank Mrs Unni Ronning and Ms Marianne Isaksen for their expert technical assistance and Mrs Eva Skaar for typing the manuscript.

References

1. Povlsen CO and Jacobsen GK: Chemotherapy of a human malignant melanoma in the nude mouse. Cancer Res 1975 (35): 2790–2796
2. Fodstad Ø, Olsnes S and Pihl A: Inhibitory effect of abrin and ricin on the growth of transplantable murine tumors and of abrin on human cancers in nude mice. Cancer Res 1977 (37): 4559–4567
3. Fogh J, Fogh JM and Orfeo T: One hundred and twenty-seven cultured human tumor cell lines producing tumors in nude mice. J Natl Cancer Inst 1977 (59): 221–226
4. Giovanella BC, Stehlin JS, Fogh J and Sharkey FE: Serial transplantation of human malignant tumors in nude mice and their use in experimental chemotherapy. In: Houchens D and Ovejera T (eds) Proceedings of the sympo-

sium on the use of athymic (nude) mice in cancer research. New York: Gustav Fischer 1978: 163-179
5. Bellet RE, Danna V, Mastrangelo MJ and Berd D: Evaluation of a "nude" mouse-human tumor panel as a predictive secondary screen for cancer chemotherapeutic agents. J Natl Cancer Inst 1979 (63): 1185-1188
6. Fodstad Ø, Aass N and Pihl A: Response to chemotherapy of human malignant melanoma xenografts in athymic nude mice. Int J Cancer 1980 (25): 453-458
7. Sordat B, Fritsche R, Mach P, Carrel S, Ozello L and Cerotini JC: Morphological and functional evaluation of human solid tumors serially transplanted in nude mice. In: Rygaard J and Povlsen CO (eds): Proceedings of the first international workshop on nude mice. Stuttgart: Gustav Fischer 1974: 269-277
8. Giovanella BC, Stehlin JS Jr, Shephard RC and Williams LJ: Correlation between response to chemotherapy of human tumors in patients and in nude mice. Cancer 1983 (52): 1146-1152
9. Fiebig HH, Schuchhardt C, Henss H, Fiedler L and Löhr GW: Comparison of tumor response in nude mice and in the patients. Behring Inst Mitt 1984 (74): 343-352
10. Fodstad Ø, Aamdal S, Pihl A and Boyd MR: Activity of mitozolomide (NSC 353451), a new imidazotetrazine, against xenografts from human melanomas, sarcomas, lung and colon carcinomas. Cancer Res 1985 (45): 1778-1786
11. Gundersen S, Aamdal S and Fodstad Ø: Mitozolomide (NSC 353451), a new active drug in the treatment of malignant melanoma. Phase II trial in patients with advanced disease. Br J Cancer 1987 (55): 433-435
12. Fodstad Ø: Representativity of xenografts for clinical cancer. Tumor and host characteristics as variables of tumor take rate. In: Winograd B, Peckham MJ and Pinedo HM (eds) Human tumour xenografts in anticancer drug development. ESO Monographs. Heidelberg: Springer 1988: 15-21
13. Steel GG and Stephens TC: The relation of cell kinetics to cancer chemotherapy. In: Adolphe M (ed) Advances in pharmacology and therapeutics. Oxford: Pergamon Press 1978 (10): 137-145
14. Shoemaker RH, McLemore TL, Abbott BJ, Fine DL, Gorelik E, Mayo JG, Fodstad Ø and Boyd MR: Human tumor xenograft models for use with an in-vitro based, disease-oriented antitumor drug screening program. In: Winograd B, Peckham MJ and Pinedo HM (eds Human tumour xenografts in anticancer drug development. ESO Monographs. Heidelberg: Springer 1988: 115-122
15. Olsnes S and Pihl A: Toxic lectins and related proteins. In: Van Heyningen S and Cohen P (eds) The molecular action of toxins and viruses. Amsterdam: Elsevier 1982: 51-105
16. Fodstad Ø, Kvalheim G, Godal A, Lotsberg J, Aamdal S, Host H and Pihl A: Phase I study of the plant protein ricin. Cancer Res 1984 (44): 862-865
17. Boven E, Nauta MM, Schluper HMM, Elferink F, Van der Vijgh WJF and Pinedo HM: Secondary screening of platinum compounds in human ovarian cancer xenografts in nude mice. Eur J Cancer Clin Oncol 1985 (21): 1253-1260
18. Stevens MFG, Hickman JA, Stone R, Gibson NW, Baig G, Lunt E and Newton CG: Antitumor imidazotetrazinones. 1. Synthesis and chemistry of 8-carbamoyl-3-(-2-chloroethyl) imidazo 5,1-d-1,2,3,5-tetrazin-4(3H)one, a novel broad-spectrum antitumor agent. J Med Chem 1984 (27): 196-201
19. Fodstad Ø, Aass N and Pihl A: Assessment of tumor growth and of response to chemotherapy of human melanomas in athymic, nude mice. Br J Cancer 1980 (41, suppl IV): 146-149
20. Nowak K, Peckham MJ and Steel GG: Variation in response of xenografts of colo-rectal carcinoma to chemotherapy. Br J Cancer 1978 (37): 576-584

IV. Preclinical Evaluation of Analogs of Known Anticancer Agents

Screening of Analogs in Human Ovarian Cancer Xenografts

E. BOVEN

Department of Oncology, Free University Hospital, De Boelelaan 1117, 1081 HV Amsterdam, The Netherlands

Abstract

Human ovarian cancer lines were used to compare the efficacy of platinum compounds with the efficacy of cisplatin and that of trimelamol with hexamethylmelamine at maximum tolerated doses. As for cisplatin analogs tested in two lines, carboplatin was shown to be similar in efficacy to cisplatin, followed by iproplatin, JM-40 and TNO-38; tetraplatin and spiroplatin were considered less active. Experimental data correlated well with clinical findings in ovarian cancer currently available for cisplatin, carboplatin, iproplatin and spiroplatin. The slightly lower efficacy of TNO-38 as compared to the parent drug was reconfirmed in 5 other human ovarian cancer lines.

The activity of trimelamol was similar to or higher than that of hexamethylmelamine in 8/9 lines studied. The human ovarian cancer model seems to have considerable value in the selection of analogs with potential activity in this malignancy.

Introduction

The aim of synthesizing analogs of conventional antitumor agents is generally to increase the therapeutic index, either by reducing the side effects or by increasing the cytotoxicity. Another important reason to generate analogs may be the selection of a new compound without cross-resistance to the parent drug.

Cisplatin is active in a variety of cancers including ovarian cancer (1). A series of analogs have subsequently been evaluated in preclinical studies in order to reduce nephrotoxicity; two of these, carboplatin and iproplatin, have recently entered clinical trials (2, 3).

Hexamethylmelamine, another compound with activity in ovarian cancer (4, 5), has to be administered orally because of its poor solubility in water. Plasma drug levels achieved are variable, due to individual differences in absorption and to a poor patient compliance because of gastrointestinal toxicity. Synthetic research has led to trimelamol, a water-soluble analog, which has been developed and tested in a phase I clinical trial (6).

Comparative studies of parent drugs and their derived products in a series of human tumor xenografts may give better insight in the potential activity of analogs in a particular malignancy in the clinical situation. For this reason we have investigated various platinum compounds in human ovarian cancer lines. These compounds were selected because in murine tumor systems each had an overall efficacy resembling that of cisplatin. We also studied the activity of trimelamol in our lines and compared the anti-tumor effects with those obtained with hexamethylmelamine.

Materials and Methods

For the use of the animals, tumor lines, procedure of transplantation and evaluation of treatment reference is made to the papers "Biology of human ovarian cancer xenografts" and "Conventional agents in human ovarian cancer xenografts" chapter I and III of this Monograph.

Treatment

Cisplatin and analogs were injected as single agents i.v. weekly for 2 or 3 weeks at doses approximating the maximum tolerated dose (MTD). Hexamethylmelamine and trimelamol were injected i.p. at MTD according to a q2d$\times$4 and q2d$\times$5 schedule, respectively. Drugs and doses are indicated in Table 1. The preparation of drugs has been described earlier (7-9).

Statistics

Antitumor activity of parent drugs and analogs was evaluated by Student's t test.

Table 1. Administered doses of cisplatin, hexamethylmelamine and their analogs

Drug	Chemical name	Dose (mg/kg)
Cisplatin	cis-diamminedichloroplatinum (II)	4-5
Carboplatin	cis-diammine-1,1-cyclobutane-dicarboxylatoplatinum (II)	60
JM-40	cis-ethylenediaminoplatinum(II)-malonate	40
Iproplatin	cis-dichlorobis(isopropylamine)-trans-dihydroxyplatinum(IV)	40
Spiroplatin	(cis-1,1-bis(aminomethyl)-cyclohexane)platinum(II)sulphate	3, 2, 2
TNO-25	(1,1-bis(aminomethyl)cyclohexane)-(2-ethylmalonato)platinum(II)	80
TNO-26	(1,1-bis(aminomethyl)cyclobutane)-dihydroxydichloroplatinum(IV)	20, 10, 10
TNO-27	(1,1-bis(aminomethyl)cyclohexane)-dichlorosulphatoplatinum(IV)	6
TNO-32	diaqua(1,1-bis(aminomethyl)-cyclohexane)platinum(II)dinitrate	3.5
TNO-38	(1,1-bis(aminomethyl)-cyclohexane)oxalatoplatinum(II)	7-8
Tetraplatin	tetrachloro(d,1-trans)1,2-diamino-cyclohexaneplatinum(IV)	5
Hexamethyl-melanime	hexamethylmelamine	150
Trimelamol	N2,N4,N6-trihydroxymethyl-N2,N4,N6-trimethylmelamine	120

Results

Platinum compounds were first studied for their activity in Ov. Pe and MRI-H-207, tumor lines which are known for their difference in sensitivity to cisplatin. Drugs were administered three times weekly in Ov. Pe and twice weekly in MRI-H-207 and the antitumor effects are summarized in Table 2. Compared to cisplatin the most active drugs in Ov. Pe were carboplatin, TNO-26, TNO-27 and tetraplatin. In MRI-H-207 cisplatin could achieve 100% growth inhibition without regrowth of tumors. Complete remissions were also observed with carboplatin, JM-40, iproplatin and TNO-38. The slightly lower efficacy of TNO-38 compared to the parent compound was confirmed in 5 other human ovarian cancer lines (Table 3). Although the efficacy of tetraplatin was similar to that of cisplatin in Ov. Pe and Ov. He, the former drug was less effective in Ov. Me and could not achieve 100% growth inhibition in MRI-H-207.

Nine human ovarian cancer lines were investigated for their sensitivity to hexamethylmelamine and trimelamol (Table 4). Except for Ov. He, the efficacy of trimelamol was similar to or higher than that obtained with the parent compound. Moreover, trimelamol could achieve 100% growth inhibition in 3/9 lines, while with hexamethylmelamine this effect was observed in MRI-H-207 only.

Table 2. Growth inhibition (%) by platinum compounds in human ovarian cancer lines

Drug	Ov. Pe	MRI-H-207
Cisplatin	55[a]	100[a]
Carboplatin	53[a]	100[a]
JM-40	24	100[a]
Iproplatin	34	100[a]
Spiroplatin	0	94[a]
TNO-25	13	96[a]
TNO-26	57[a]	99[a]
TNO-27	51[a]	91[a]
TNO-32	44	58[a]
TNO-38	15	100[a]
Tetraplatin	52[a]	99[a]

[a] $p \leq 0.05$.

Table 3. Growth inhibition (%) by cisplatin, TNO-38 and tetraplatin in human ovarian cancer lines

Tumor line	Cisplatin	TNO-38	Tetraplatin
Ov. Pe	55[a]	15	52[a]
MRI-H-207	100[a]	100[a]	99[a]
Ov. He	56[a]	47[a]	53[a]
Ov. Gl	77[a]	61[a]	ND
Ov. Ri(C)	92[a]	52[a]	ND
FKo	14	23	ND
Ov. Me	64[a]	37	48

[a] $p < 0.05$, ND = not done.

Table 4. Growth inhibition (%) by hexamethylmelamine and trimelamol in human ovarian cancer lines

Tumor line	Hexamethylmelamine	Trimelamol
Ov. Pe	60[a]	81[a]
MRI-H-207	100[a]	100[a]
Ov. He	73[a]	42[a]
Ov. Gl	87[a]	93[a]
Ov. Ri(C)	61[a]	98[a]
FMa	98[a]	100[a]
FKo	35[a]	92[a]
A 2780S	80[a]	79[a]
Ov. Me	99[a]	100[a]

[a] $p < 0.05$.

Discussion

Using human ovarian cancer lines varying in histology, growth rate and chemosensitivity, we compared the differences in efficacy between analogs and the parent compounds cisplatin and hexamethylmelamine. Our model appeared to yield reliable data since a platinum compound with lower activity than cisplatin consistently resulted in reduced efficacy. On the other hand, with trimelamol we consistently observed superior antitumor activity as compared to hexamethylmelamine, with the exception of one tumor line.

In clinical studies carboplatin and iproplatin have been shown to be active in patients with ovarian cancer (2, 3). The nephrotoxic compound spiroplatin, however, has proved to be ineffective in patients (10). Clinical findings correlate well with data in human ovarian cancer xenografts. We demonstrated that the cisplatin-sensitive MRI-H-207 tumor line was equally sensitive to carboplatin and iproplatin and less sensitive to spiroplatin. The TNO-compounds, JM-40 and tetraplatin, also possessed antitumor activities in MRI-H-207, but with regard to results in other tumor lines, such as Ov.Pe, none appeared to be superior to cisplatin.

There are several advantages in substituting trimelamol for hexamethylmelamine in the clinic. Hexamethylmelamine requires metabolic activation by a liver microsomal fraction in order to exert its cytotoxic action (11). The direct administration of an active metabolite (N-methylolmelamines), such as trimelamol, would obviously facilitate drug action (12). Unlike hexamethylmelamine, trimelamol can be administered i.v., which should optimize drug levels. Finally, the superior efficacy of trimelamol as compared to hexamethylmelamine in our human ovarian cancer lines indicates an increased therapeutic index for this analog. Preliminary results of trimelamol in a Phase I trial suggest that the compound has antitumor activity in ovarian cancer patients (6).

References

1. Richardson GS, Scully RE, Nikrui N and Nelson JH: Common epithelial cancer of the ovary (second of two parts). N Engl J Med 1985 (312): 474-483
2. Evans BD, Raju KS, Calvert AH, Harland SJ and Wiltshaw E: Phase II study of JM8, a new platinum analog, in advanced ovarian carcinoma. Cancer Treat Rep 1983 (67): 997-1000
3. Creaven PJ, Madajevicz S, Pendyala L, Mittelman A, Pontes E, Spaulding M, Arbuck S and Solomon J: Phase I clinical trial of cis-dichloro-trans-dihydroxy-bis-isopropylamine platinum (IV) (CHIP). Cancer Treat Rep 1983 (67): 795-800
4. Johnson BL, Fisher RI, Bender RA, DeVita VT, Chabner BA and Young RC: Hexamethylmelamine in alkylating agent-resistant ovarian carcinoma. Cancer 1978 (42): 2157-2161
5. Bonomi PD, Mladineo J, Morrin B, Wilbanks G and Slayton RE: Phase II trial of hexamethylmelamine in ovarian carcinoma resistant to alkylating agents. Cancer Treat Rep 1979 (63): 137-138
6. Judson IR, Rutty CJ, Abel G, Gumbrell L, Harrap KR and Calvert AH: Report on the phase I trial of N^2,N^4,N^6-trihydroxymethyl-N^2,N^4,N^6-trimethylmelamine (trimelamol). Br J Cancer 1985 (52): 466
7. Boven E, Van der Vijgh WJF, Nauta MM, Schlueper HMM and Pinedo HM: Comparative acivity and distribution studies of five platinum analogues in nude mice bearing human ovarian cancer xenografts. Cancer Res 1985 (45): 86-90
8. Boven E, Nauta MM, Schlueper HMM, Elferink F, Van der Vijgh WJF and Pinedo HM: Secondary screening of platinum compounds in human ovarian cancer xenografts in nude mice. Eur J Cancer Clin Oncol 1985 (21): 1253-1260
9. Boven E, Nauta MM, Schlueper HMM, Erkelens CAM and Pinedo HM: Superior efficacy of trimelamol to hexamethylmelamine in human ovarian cancer xenografts. Cancer Chemother Pharmacol 1986 (18): 124-128
10. Vermorken JB, Ten Bokkel Huinink WW, McVie JG, Van der Vijgh WJF and Pinedo HM: Clinical experience with 1,1-diaminomethylcyclohexane(sulphato)platinum(II) (TNO-6). In: Hacker MP, Douple EB and Krakoff IH (eds) Platinum coordination complexes in cancer chemotherapy. Boston: Martinus Nijhoff Publishing: 330-343
11. Ames MM, Sanders ME and Tiede WS: Metabolic activation of hexamethylmelamine and pentamethylmelamine by liver microsomal preparations. Life Sciences 1981 (29): 1591-1598
12. Rutty CJ, Newell DR, Muindi JRF and Harrap KR: The comparative pharmacokinetics of pentamethylmelamine in man, rat and mouse. Cancer Chemother Pharmacol 1982 (8): 105-111

Predictability of Clinical Response of Several Human and Murine Tumor Models to Four Anthracycline Derivatives

G. Pezzoni and F. C. Giuliani

Farmitalia Carlo Erba, R&D/Oncology, Research Center, Nerviano (MI), Italy

Abstract

In order to establish the predictability and the usefulness of human and murine tumor models for the screening of new anthracycline derivatives, we carried out a retrospective analysis of experimental and clinical results of 4 anthracycline derivatives: doxorubicin, epirubicin, esorubicin and idarubicin.

Due to the different endpoints of response and criteria for the evaluation of antitumor activity in patients and experimental models, this analysis can only be qualitative.

We found that experimental mammary tumors (xenografts and murine) can be considered to be the most predictive models for patient chemosensitivity to anthracycline derivatives. Few data are available to ascertain the predictability of lung tumor models.

Our data indicate that colorectal carcinoma models can be used to predict clinical inactivity of anthracycline derivatives.

Experimental melanomas appear useless in the screening of new derivatives, since these models give rise to false positive drugs.

Introduction

Most clinically active chemotherapeutic agents have been developed on the basis of activity against transplantable or spontaneous murine tumors. Recently, new preclinical models have been utilized to evaluate candidate antitumor compounds, and particular interest has been focused upon the nude mouse human tumor system (1, 2). It is claimed that this model is appropriate to test the *in vivo* chemosensitivity of human tumors, which maintain their morphological and biochemical characteristics after transplantation into nude mice (3, 4).

Many retrospective studies (5-9) have indicated that there is a correlation between xenograft chemosensitivity and the sensitivity of the same tumors in patients. This has been reported for different clinically active drugs such as 5-fluorouracil, mitomycin C, BCNU, cyclophosphamide, doxorubicin and recently for new platinum analogs (10). For example, Giovanella (11) evaluated the chemosensitivity of 14 melanomas, 14 colorectal carcinomas and 14 breast carcinomas, and reported that the results indicated that heterotransplantation of human tumors was a valid predictive system for testing new anticancer compounds.

In recent years, our research group has been involved in the search for new anthracycline analogs, offering improvements over the parent drugs doxorubicin (DX) and daunorubicin (DNR).

We present a retrospective evaluation of clinical efficacy (overall patients' response) after single-agent treatment compared with activity on experimental models of four anthracyclines: DX, EPI (4-epidoxorubicin) (12), ESO (4′-deoxydoxorubicin) (13) and IDA (4′-demethoxydaunorubicin) (14). This assessment can obviously only be qualitative, since there are major differences between criteria for evaluation of antitumor activity in patients and experimental models. In animals the usual endpoint of response is tumor growth inhibition, whereas in patients mere slowing of growth is not reliably detectable and any growth in excess of 25% beyond the baseline is considered progression. Moreover, in most experimental models, regression of sizeable established tumors is not readily attainable, and that is the usual goal of human chemotherapy. Despite these differences encountered in any discussion of the predictability of experimental tumor systems, our retrospective analysis may nevertheless be helpful in deciding which models are most suitable for clinical predictability in the development of new anthracycline analogs.

Materials and Methods

Animals

Balb/c/nu/nu mice 8 to 12 weeks old were used for experimental xenografting of human tumors; see reference 15 for maintenance and breeding conditions.

Inbred female C3H/He, C57BL/6 and Balb/c adult mice were used for experimental murine tumors. These mice were supplied by Charles River Italia (Calco - Como, Italy).

Tumors

Human tumors: Please consult references 15 and 16 for a list of tumors tested.
Xenografts were established by inoculation of fresh tumor tissue from patients into nude mice as described (16).

Murine tumors: mammary carcinoma (20×10^6 cells/mouse) from a third generation tumor was inoculated s.c. in C3H female mice.
B16 melanoma (25×10^6 cells/mouse) was transplanted s.c. in C57BL/6 mice.
Colon 38 and Colon 26 were transplanted s.c. (fragments) in C57BL/6 mice and Balb/c mice, respectively.

Results

Table 1 shows the qualitative comparison of antitumor activity of 4 anthracycline derivatives between tumors implanted in nude mice and the clinical efficacy in patients with the same tumor types. Definitions of activity are reported in the Table legend.

DX and EPI, tested against 6 human tumor types, were the most active drugs. DX and EPI were active in breast carcinoma and lung SCC, both in the experimental model (15, 16) and clinically (17, 18); in colorectal cancer DX was inactive in both cases (19), whereas EPI was inactive in patients (17, 18) but showed borderline activity in xenograft models (20). ESO was active in breast xenografts, showing a correlation with clinical results (17); it also showed a certain correlation, in terms of inactivity, for lung. However, the greatest difference for this doxorubicin analog was observed for colorectal cancers: 5/7 human colorectal tumors xenografted in nude mice were responsive to ESO (20), while in colorectal clinical trials ESO was inactive (17, 18). IDA is mainly active in clinical use in leukemias and lymphomas (17); its clinical antitumor activity in solid tumors is not yet well defined, but it has recently been reported to be active in breast carcinoma, thus confirming its activity in the same tumor in the xenograft system. No correlation was observed for melanoma: all 4 drugs were active in the experimental system but inactive in clinical use (17).

The data reported in Table 2 show the activity of the 4 anthracycline derivatives in breast and colorectal carcinoma and melanoma in murine tumors and in human tumors implanted in nude mice, as compared with clinical efficacy.

All drugs were active in breast carcinoma, as confirmed clinically (17); colorectal murine tumors were responsive to all drugs but colorectal tumors were not responsive in patients; this with the exception of IDA, which has not yet been evaluated in the clinic. Responses differed in human colorectal carcinoma implanted in nude mice: DX was inactive (19), ESO active (20) and EPI and IDA showed borderline activity (20–21). As far as melanoma is concerned,

Table 1. Qualitative comparison of antitumour activity between human tumor xenografts and clinical results for DX, EPI, ESO and IDA on different human solid tumors

Tumor	DX		EPI		ESO		IDA	
	Exp.[a]	Clin.[b]	Exp.	Clin.	Exp.	Clin.	Exp.	Clin.
Breast carcinoma (2–4)[d]	+++	++[c]	++	++[c]	+++	+[c]	++	+[c]
Rectum carcinoma (2–5)	−	−[c]	−	+/−	+	−	+	
Colon carcinoma	−	−[c]	+	−	+++	−	+	
Lung NSCC (2)	++	+/−	++	−	−	−[c]		
Lung SCC (1)	+++	+[c]	+++	+[c]				
Melanoma (2–4)	++	−	+++	−	+	−	−	−[c]

[a] Experimental results: +++ tumor growth inhibition > 80%; ++ more than 50% but less than 80%; + < 50%; − no response.
[b] Clinical response rate (CR + PR): ++ > 50%; + more than 20% but less than 50%; − no response. Data from refs. 17 and 18.
[c] correlation.
[d] number of tumor xenografts tested.

Table 2. Qualitative comparison between preclinical and clinical results for DX, EPI, ESO and IDA on different tumor models

Tumor	DX			EPI			ESO			IDA		
	Mur[a]	Xen[b]	Clin[c]	Mur	Xen	Clin	Mur	Xen	Clin	Mur	Xen	Clin
Breast carcinoma	+	+	+	+	+	+	+/−	+	+	+/−	+	+
Colorectal carcinoma	+	−	−	+	+/−	−	+	+	−	+	+/−	
Melanoma	+	+	−	+	+	−	+/−	+/−	+		−	−

[a] Antitumor activity on murine tumors.
[b] Antitumor activity on human tumors xenografted in nude mice.
[c] Clinical response.

neither the murine tumor (B16) nor the human xenografted tumors are predictive tests for clinical efficacy: drugs were active in both models but inactive in patients (17–18).

Discussion

Are the experimental tumor models, murine and human, truly considered representative of the populations to which the clinical activity of new anthracycline derivatives is oriented? This is the question we set out to answer.

In the present retrospective evaluation we found that the most active drugs in experimental models, DX and EPI, maintained this activity in patients, particularly for breast carcinoma, and except for melanoma. ESO was less active than DX and EPI, showing a narrower spectrum of activity in experimental tumor models and in patients; it was more active on experimental colorectal tumors but this was not maintained in patients. This lack of predictability of experimental colorectal tumors in relation to ESO could be due to different reasons: small number of tumors tested, greater drug tolerance and different drug pharmacokinetics or metabolic properties in mice and humans.

We found that the experimental mammary tumors used *in vivo* in the development of new anthracycline analogs were the most predictive models for patient chemosensitivity. The clinical predictability of experimental lung carcinomas probably depends on histiotype sensitivity; we have seen that small cell lung carcinoma xenografts can be as sensitive as in patients to the same analogs.

Finally, experimental melanomas are not representative for clinical predictability of new anthracyclines. No correlation was obtained between preclinical and clinical results for any of the analogs tested. In order to explain the lack of predictability with melanoma we suggest the following differences between man and mice: different pattern of tumor heterogeneity, tumor cell kinetics, vascularization, different ratio between tumor mass and host body mass and different metastatic pattern.

In conclusion, we consider experimental breast cancer models to be valid for anthracycline screening. Our data indicate that colorectal carcinoma models can be used to predict clinical inactivity, whereas experimental melanomas appear useless, since these models give rise to false positive drugs.

Acknowledgements

We thank Drs. Maria Grandi and Charles W. Young for helpful discussions and advice.

References

1. Houchens DP and Ovejera AA: Model for human tumor therapy in nude mice. Immune-deficient Animals. Int Workshop on Immune-deficient Animals in Exp Res. Chexbres 1982: 364-369. Karger, Basel 1984
2. Ovejera AA and Houchens DP: Human tumor xenografts in athymic nude mice as a preclinical screen for anticancer agents. Sem in Oncol 1981 (8): 386-393
3. Povlsen A and Rygaard J: Growth of human tumors in the nude mouse. In: Bloom BB, David JR (eds) In vitro methods in cell mediated and tumor immunity. Academic Press, New York 1976 pp 701-711
4. Povlsen CD and Rygaard J: Heterotransplantation of human adenocarcinoma of the colon and rectum to the mutant nude mouse. A study of nine consecutive transplantations. Acta Path Microbiol Scand 1971 (79, Section A): 159-169
5. Shorthouse AJ, Peckham MJ, Smyth JF and Steel GG: The therapeutic response of bronchial carcinoma xenografts: a direct patient-xenograft comparison. Br J Cancer 1978 (41 Suppl IV): 142-145
6. Inaha H, Tashiro T, Kobayashi T, Fujimoto S, Sakurai Y, Mauro K, Ohnishi Y, Ueyama Y and Nomura T: Evaluation of response rates to various antitumor agents of human gastric tumors implanted in nude mouse. Jap J Cancer Res 1986 (77): 190-196
7. UICC Study Group on Chemosensitivity Testing of Human Tumors. Problems - Applications - Future Prospects. Int J Cancer 1986 (37): 1-5
8. Fujita M, Hayata S and Taguchi T: Relationship of chemotherapy on human cancer xenografts in nude mice to

clinical response in donor patient. J of Surg Oncol 1980 (15): 211-219
9. Steel GG, Courtenay VD and Peckham MJ: The response to chemotherapy of a variety of human tumor xenografts. Br J Cancer 1983 (47): 1-13
10. Boven E, Vijgh W, Nauta M, Schluper M and Pinedo HM: Comparative activity and distribution studies of five platinum analogues in nude mice bearing human ovarian carcinoma xenografts. Cancer Res 1985 (45): 86-90
11. Giovanella BC, Stehlin JS, Shepard RC and Williams LY: Correlation between response to chemotherapy of human tumors in patients and in nude mice. Cancer 1983 (52): 1146-1152
12. Arcamone F, Cassinelli G, Penco S, Vicario GP, Vigevani A: Chemistry of Epirubicin. In: G. Bonadonna (ed) Advances in Anthracyclines Chemotherapy: Epirubicin. Masson, Italy
13. Arcamone F, Penco S, Redaelli S, Hanessian S: Synthesis and antitumor activity of 4′-deoxydaunorubicin and 4′-deoxyadriamycin. J Med Chem 1976 (19): 1424-1425
14. Arcamone F, Bernardi L, Giardino P, Patelli B, Di Marco A, Casazza AM, Pratesi G, Reggiani P: Synthesis and antitumor activity of 4-demethoxydaunoribicin, 9-diepidaunorubicin and their Beta anomers. Cancer Treat Rep 1976 (60): 829-836
15. Giuliani FC, Zirvi K, Kaplan NO: Therapeutic response of human tumor xenografts in athymic mice to doxorubicin. Cancer Res 1981 (41): 325-335
16. Giuliani FC, Coirin AK, Rice R, Kaplan NO: Effect of 4′-doxorubicin analogs on heterotransplantation of human tumors in congenitally athymic mice. Cancer Treat Rep 1981 (65): 1063-1075
17. Hurteloup P, Ganzina F: Clinical studies with new anthracyclines: Epirubicin, Idarubicin, Esorubicin. Drug Exp Clin Res 1986 (XII): 233-246
18. Martoni A, Giovannini M, Tomasi L et al: A phase-II clinical trial of 4′-epidoxorubicin in advanced solid tumors. Cancer Chemother Pharmacol 1984 (12): 179-182
19. Giuliani FC: Valutazione dell'activita' antineoplastica di nuovi farmaci: tumori sperimentali e xenotrapiantati di tumori umani in topi atimici. Argomenti di Oncologia 1983 (4): 1-10
20. Giuliani FC and Kaplan NO: New doxorubicin analog active against doxorubicin-resistant colon tumor xenografts in the nude mouse. Cancer Res. 1980 (40): 4682-4687
21. Giuliani FC and Kaplan NO: The use of athymic mice in the screening of new anthracycline derivatives. Immune-deficient Animals. 4th Int Workshop on Immune-deficient Animals in Exp Res. Chexbres 1982: 374-374. Karger, Basle 1984

V. New Drugs

Activity of Six New Drugs in Head and Neck Cancer Xenografts

B. J. M. Braakhuis, A. Leyva[1], H. M. Pinedo[1], and G. B. Snow

Departments of Otolaryngology and Oncology[1], Free University Hospital, De Boelelaan 1117, 1081 HV Amsterdam, The Netherlands

Abstract

The nude mouse xenograft model was used to test 6 new agents for activity against head and neck cancer. Animals were treated intraperitoneally at maximum tolerated doses. 5-Aza-2′-deoxycytidine (5-aza-dCyd), a nucleoside analogue, was active in 1, moderately active in another and inactive in 3 tumour lines. Activity was dose-schedule dependent, indicating that new drugs ought to be tested in at least 2 dose-schedules. Trimetrexate, a new folate antagonist, was tested in 3 tumour lines insensitive to methotrexate, and was shown to be inactive. The activity of 5′-deoxy-5-fluorouridine (5′dFUR), a prodrug of 5-fluorouracil (5-FU) was compared with that of 5-FU. In 1 line a better antitumour effect was seen with 5′dFUR than with 5-FU. In another line, insensitive to 5-FU, only a minimal response to 5′dFUR was detected. Three platinum analogues, TNO-6, JM-8 and JM-40 were tested in cisplatin insensitive lines. No superior effect of the analogues as compared to cisplatin was observed. In single head and neck tumour lines 5′dFUR and 5′aza-dCyd have a better antitumour effect than conventional drugs; this makes these drugs of potential value for patients with head and neck cancer.

Introduction

There is an urgent need for new chemotherapeutic agents to treat patients with head and neck cancer. With the conventional drugs remissions can be obtained in 20–40% of patients, but survival time is usually not or only minimally increased (1).

The nude mouse model seems suitable for the evaluation of new anticancer drugs. Xenografted tumours respond to agents that are active in the clinic (2). At this moment it is too early to state that the results of the model always correlate with those obtained in patients in various tumour types. For head and neck cancer more conventional drugs should be evaluated in various lines (3).

Based on the results obtained with the model so far and the urgent need for new drugs for head and neck cancer, we decided to test some new drugs in head and neck cancer xenografts. Our programme was limited to the testing of drugs that were studied in phase I trials. Our results might help in the selection of compounds for subsequent phase II trials in patients with head and neck cancer.

Materials and Methods

Animals and Tumours

Experiments were carried out in a similar way as described previously (3). In addition to the lines reported, we used the head and neck tumour line HNX-Hep-2, established from a cell line by subcutaneous injection of 3×10^6 cells. Antitumour effect was expressed as growth delay factor (GDF), as described previously (3).

Drugs

Cisplatin (CDDP, platinol), JM-40, TNO-6 (spiroplatin) and JM-8 (carboplatin) were supplied by Bristol Myers International, New York, USA. 5-Fluorouracil (5-FU) and 5′-deoxy-5-fluorouridine (5′dFUR, doxifluridine) were donated by Hoffmann-La Roche, Mijdrecht, The Netherlands, 5-aza-2′-deoxycytidine (5-aza-dCyd) by Pharmachemie, Haarlem, The Netherlands and trimetrexate (TMQ) by Warner Lambert, Ann Arbor, USA. Methotrexate was obtained from Lederle, Etten-Leur, The Netherlands.

Results and Discussion

Three new antimetabolites were tested in 5 head and neck tumour lines and were, if possible, compared with related conventional agents. 5-Aza-dCyd, a nucleoside analogue, is an active antileukaemic agent in animal models (4) and children (5). This

Table 1. Activity of 5-aza-2′-deoxycytidine in head and neck cancer xenografts

Line	Dose (mg/kg)	Injections on day	GDF[a]	Regression[b]
HNX-DU	2	0, 7	2.2	0/7
HNX-DU	2	0, 4, 8	3.6	1/6
HNX-DU	0.25	0–4, 7–11	0.2	0/7
HNX-DU	0.1	0–4[c]	0.2	0/4
HNX-Hep-2	2	0, 4, 8	0.5	0/9
HNX-KE	2	0, 4, 8	0.1	0/5
HNX-KE	0.1	0–4[c]	0.0	0/7
HNX-FA	2	0, 4	0.9	0/7
HNX-FR	2	0, 4	0.5	0/6

[a] growth delay factor; see: materials and methods of ref. 3. Due to the fact that tumours sometimes regressed completely median values of tumour doubling times were used for the computation of GDF.
[b] no. regressed / no. treated tumours.
[c] q4hrs × 3.

drug is now being evaluated in Europe in phase I and II trials in patients with solid tumours and leukaemia. 5-Aza-dCyd had a substantial antitumour effect in 1 tumour line, HNX-DU, derived from an undifferentiated carcinoma of the hypopharynx. With a schedule of 2 mg/kg on days 0, 4 and 8 a GDF of 3.6 was seen and 1 out of 6 tumours regressed completely after treatment (Table 1). A minimal effect was noted when the drug was given at a low dose daily for a longer period of time. When compared with other drugs it appeared that 5-aza-dCyd and CDDP (given weekly) were the best agents in this line (Tables 1, 2 and 3). 5-Aza-dCyd was also tested in 4 other tumour lines; in 1 line a moderate response was observed and in the 3 other lines the drug was inactive (Table 1).

5′dFUR is a structural analogue of 5-FU. In line HNX-KE, a poorly differentiated squamous cell carcinoma of the larynx, treatment with 5′dFUR resulted in a better antitumour effect than was obtained with 5-FU, with GDF-values of 2.1 and 0.8, respectively (Table 2). In another 5-FU insensitive line, the therapeutic effect of 5-FU could not be improved. The reason why 5′dFUR can be a better drug than 5-FU, is not completely understood. The activity of pyrimidine nucleoside phosphorylase, an enzyme needed for the conversion of 5′-dFUR to 5-FU might be increased in tumour tissue. Also the mechanism of action may vary between the 2 drugs (6).

TMQ is a new lipophilic antifolate drug, active in cell lines with defective MTX transport (7). We have tested this drug in 3 MTX-insensitive lines. With a dose schedule that proved to be active against murine L1210 *in vivo* (J. R. Bertino, personal communi-

Table 2. Activity of conventional and new antimetabolites in head and neck cancer xenografts expressed as growth delay factor (GDF) (see ref. 3)

Line	5-FU	5′dFUR	MTX	TMQ
HNX-DU	0.4	0.4	0.4	0.4
HNX-KE	0.8	2.1	0.3	0.6
HNX-W	n. t.	n. t.	0.1	0.0

Drugs were administered i.p. in the following doses and schedules: 5-FU, 100 mg/kg, q7d × 2; 5′dFUR, 1000 mg/kg, q7d × 2; MTX, 5 mg/kg qd × 5 and TMQ, 50 mg/kg, qd × 5.
n. t.: not tested.

Table 3. Activity of platinum analogues in head and neck cancer xenografts expressed as growth delay factor (GDF) (see ref 3)

Line	CDDP	TNO-6	JM-8	JM-40
HNX-DU	0.4	0.0	1.5	0.2
HNX-DU	1.8[a]			
HNX-E	0.2	0.3	n. t	n. t
HNX-FR	0.1	n. t	n. t	0.6
HNX-G	0.1	n. t	0.1	0.8
HNX-KE	0.5	n. t	0.8	n. t

Due to the fact that sometimes tumours regressed completely, median values of tumour volume doubling times were used for the computation of GDF.
Drugs were administered i.p. in the following doses and schedules: CDDP, 2 mg/kg, qd × 3; TNO-6, 1 mg/kg, qd × 3; JM-8, 20 mg/kg, qd × 3; JM-40, 10 mg/kg, qd × 3.
[a] CDDP, 4 mg/kg, q7d × 2, ⅙ of the treated tumours regressed.
n. t.: not tested.

cation) no effect was observed in these lines (Table 2). These results indicate that MTX-insensitivity in head and neck tumour lines is not likely to be caused by an impaired transport mechanism as discussed previously (3).

Three platinum analogues were tested as well and compared to CDDP. Administered in doses with comparable toxicity, no superior effect of any of the analogues is observed in 5 CDDP-insensitive lines (Table 3). These results are in agreement with those of Boven et al. (8), who found cross-resistance to CDDP with the same analogues using human ovarian tumour lines.

References

1. Hong WK and Bromer R: Chemotherapy in head and neck cancer. N Engl J Med 1983 (308): 75–79
2. Steel GG, Courtenay VD and Peckham MJ: The response to chemotherapy of a variety of human tumour xenografts. Br J Cancer 1983 (47): 1–13
3. Braakhuis BJM and Snow GB: Activity of conventional drugs in head and neck cancer xenografts. In: Winograd B,

Peckham MJ and Pinedo HM (eds) Human tumour xenografts in anticancer drug development. ESO Monographs. Heidelberg: Springer 1988: 37-40
4. Momparler RL, Bouchard J, Onetto N and Rivard GE: 5-Aza-2′-deoxycytidine therapy in patients with acute leukemia inhibits DNA methylation. Leukemia Res 1984 (8): 181-185
5. Rivard GE, Momparler RL, Demers J, Benoit P, Raymond R, Lin K and Momparler LF: Phase I study on 5-aza-2′-deoxycytidine in children with acute leukemia. Leukemia Res 1981 (5): 453-462
6. Peters GJ, Laurensse E, Leyva A, Lankelma J and Pinedo HM: Sensitivity of human, murine and rat cells to 5-fluorouracil and 5′-deoxy-5-fluorouridine in relation to drug-metabolizing enzymes. Cancer Res 1986 (46): 20-28
7. Diddens H, Niethammer D and Jackson RC: Patterns of cross-resistance to the antifolate drugs trimetrexate, metoprine, homofolate and CB3717 in human lymphoma and osteosarcoma cells resistant to methotrexate. Cancer Res 1983 (43): 5286-5292
8. Boven E, van der Vijgh WJF, Nauta MM, Schlueper HMM and Pinedo HM: Comparative activity and distribution studies of five platinum analogues in nude mice bearing human ovarian xenografts. Cancer Res 1985 (45): 86-90

Activity of Metallocene Complexes Against Human Tumours Heterotransplanted into Athymic Mice

P. KOEPF-MAIER

Institut für Anatomie, Freie Universität Berlin, Königin-Luise-Straße 15, 1000 Berlin 33, Federal Republic of Germany

Introduction

Metallocene compounds represent a new class of organometallic antitumour agents containing metals other than platinum metals. Neutral titanocene complexes of the $(C_5H_5)_2TiX_2$ type and salt-like ferricenium complexes $[(C_5H_5)_2Fe]^+X^-$ are the main representatives.

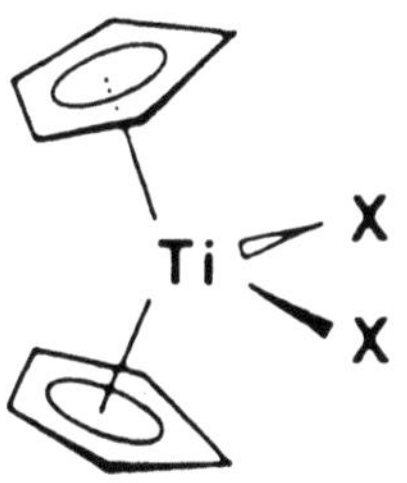

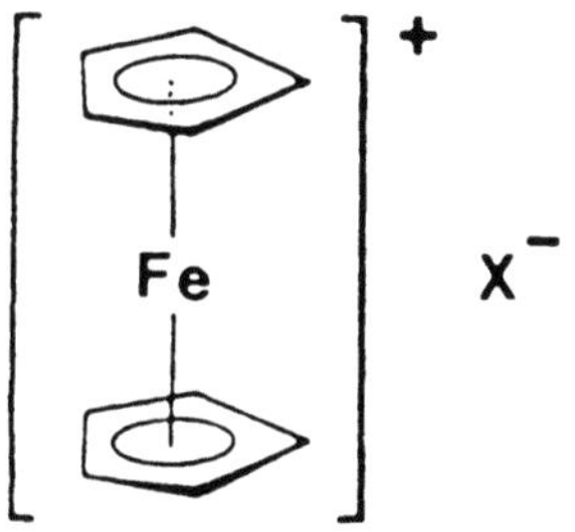

X = Cl
X = Br
X = cis-OOCCH = CHCOOH
X = p-$SC_6H_4NH_3{}^+Cl^-$

$X^- = [CCl_3COO]^- \cdot CCl_3COOH$
$X^- = [CCl_3COO]^- \cdot 2\ CCl_3COOH$
$X^- = \frac{1}{2}\ [Cl_3FeOFeCl_3]^{2-}$
$X^- = [FeCl_4]^-$
$X^- = [2,4,6\text{-}(NO_2)_3C_6H_2O]^-$

In recent years, antitumour properties of titanocene and ferricenium complexes were established against various ascitic and solid animal tumours, including Ehrlich ascites tumours (1-4), sarcoma 180 (5), melanoma B16 (6), colon 38 carcinoma (6) and Lewis lung carcinoma (7, 8). The aim of the present pilot study was to explore the activity of metallocene complexes against human tumours heterotransplanted into athymic mice.

Materials and Methods

Five human xenografts (2 colorectal adenocarcinomas, 1 breast carcinoma, 1 lung adenocarcinoma, 1 small cell lung carcinoma) were transplanted serially into nude mice. For propagation purposes, the tumours were removed from donor animals, minced mechanically, diluted with Hank's salt solution and injected subcutaneously as tumour brei into the flanks of athymic NMRI (nu/nu mice). As soon as the tumours had reached a size of 400-800 mm^3, chemotherapy was initiated. The drugs were given intraperitoneally according to Q2D × 5 or Q3D × 5 schedules. The highest doses evaluated corresponded to a LD_{10}, the lower doses applied were equivalent to ⅔ LD_{10} and ⅓ LD_{10} doses.
Two perpendicular diameters (length a, width b) of the tumours were measured every 2 or 3 days and tumour volumes were computed using the formula $v = \frac{1}{2}\ a\ b^2$. As parameters of evaluation we calculated T/C ratios in percentage. Tumour growth inhibition was quantified by calculating 100% − T/C.

Results

Titanocene complexes $(C_5H_5)_2TiX_2$ effected growth inhibition in most of the human tumours that were investigated in the present study. The results are summarized in Table 1 while some growth curves are illustrated in Figures 1 and 2.

- In the case of the adenocarcinoma derived from the colon sigmoideum, marked inhibition of tumour growth of 70-80% was brought about by titanocene dichloride and titanocene dibromide. After administration of both titanocene dihalides, tumour enlargement stagnated totally between days 19 and 27 (Fig. 1). Subsequently, a slight revival of tumour proliferation was seen. For the purpose of comparison cisplatin and 5-fluorouracil were also tested at equitoxic doses. Growth inhibition of only 10-40% was induced by these cytostatic drugs.
- When we examined antitumour activity against a rectum adenocarcinoma, only moderate activity could be demonstrated for titanocene dichloride and titanocene dibromide, resulting in tumour growth inhibition of 35-55%.

Table 1. Growth inhibitions[a] of human xenografts by titanocene complexes

	Colon adenocarcinoma	Rectum adenocarcinoma	Breast carcinoma	Lung adenocarcinoma	Lung small cell carcinoma
Control	0	0	0	0	0
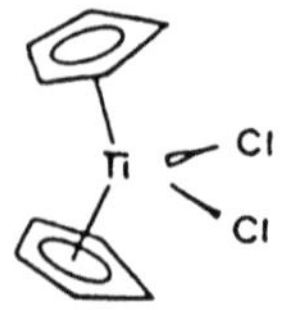	77	55	78	65	61
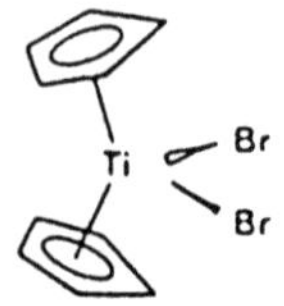	69	35	69	46	57
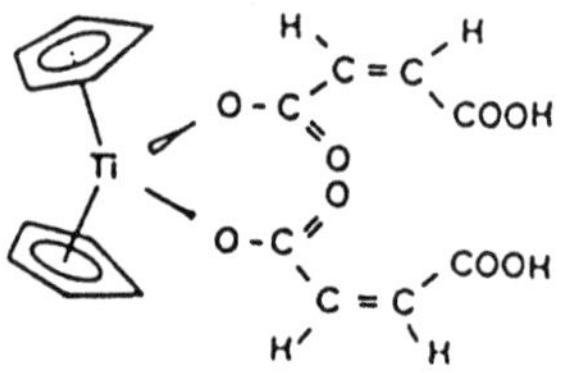	28	n.d.	n.d.	65	75
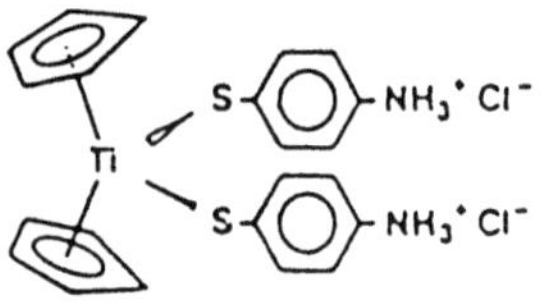	57	n.d.	n.d.	76	63

[a] Given are optimum inhibition values in % of control tumor size (calculated by 100% − T/C) 3 or 4 days after last substance application. Substance application according to Q3D × 5 or Q2D × 5. Applied doses correspond to LD_{10} regimens.

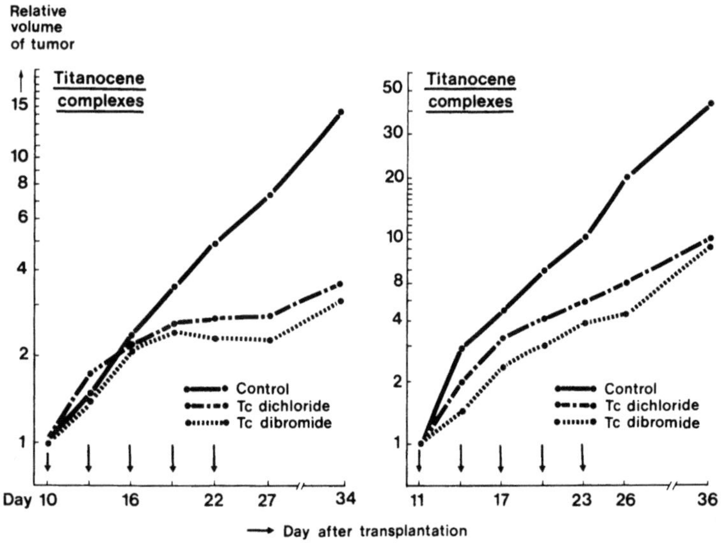

Fig. 1. Growth curves of xenografted human carcinomas after treatment with titanocene (Tc) complexes. *(left)* Colon adenocarcinoma, treated with titanocene dichloride (Q3D × 5, D = 15 mg/kg) and titanocene dibromide (Q3D × 5, D = 20 mg/kg). *(right)* Breast carcinoma, treated with titanocene dichloride (Q3D × 5, D = 10 mg/kg) and titanocene dibromide (Q3D × 5, D = 10 mg/kg)

- When titanocene compounds were administered to animals bearing rapidly proliferating human breast carcinomas, we saw again a growth inhibition of 70–80% caused by titanocene dichloride and titanocene dibromide. The growth rate of treated tumours was clearly slowed down in comparison to controls (Fig. 1). However, we did not observe any complete stagnation or absolute decrease in tumour size.
- In adenocarcinoma of the lung treatment with the dichloro, bis(hydrogenmaleinato) and bis(p-amino-thiophenolato) derivatives reduced tumour growth by 60–75% of the size of the control tumours. These values remained stable beyond the end of the treatment period (Fig. 2). There was no revival of tumour proliferation and a decrease in tumour size could be noted within 10 to 17 days after the last injection.
- In small cell lung carcinoma, only the application of maximally tolerated doses (corresponding to LD_{10} regimens) of titanocene complexes resulted in distinct growth inhibitions of 60–70%, whereas lower doses (⅓ LD_{10}, ⅔ LD_{10}) reduced tumour growth by only 0–40%. When cisplatin was administered, LD_{10} regimens were necessary to produce growth inhibition of more than 50%.

When we considered the effects induced by ferricenium compounds $[(C_5H_5)_2Fe]^+X^-$, we obtained the following findings (summarized in Table 2):

- In colon adenocarcinoma, tumour growth inhibition of more than 50% was obtained with the oxo-bridged bis(trichloroferrate), tetrachloroferrate and picrate derivatives. However, we were unable to obtain the high inhibition values of titanocene dichloride.
- In rectum carcinoma, on the other hand, ferricenium complexes exhibited a stronger antitumour activity than titanocene compounds. Growth inhibitions of 63–88% were achieved, whereby superior activity was again revealed for those ferricenium derivatives that gave the best results against colon adenocarcinoma.
- In the case of breast carcinoma, no antiproliferative effect was observed for most of the ferricenium complexes. Only the oxo-bridged bis(trichloroferrate) derivative showed marginal activity at the LD_{10} level.
- When animals bearing lung adenocarcinoma were treated with ferricenium compounds, a growth reduction of more than 50% was obtained by most of the ferricenium complexes listed in Table 2. They inhibited tumour growth during and after the treatment period, although no absolute decrease in tumour size was seen (Fig. 2).

Conclusions

The results of the present pilot study show that organometallic titanocene and ferricenium complexes exhibit antiproliferative properties against various human tumours heterotransplanted to athymic mice. Remarkably, the spectra of activity characterizing titanocene and ferricenium compounds were found not to be identical. Although the number of tumours tested in the present study was relatively small, the findings obtained indicate that metallocenes are interesting compounds for further antitumour investigations.

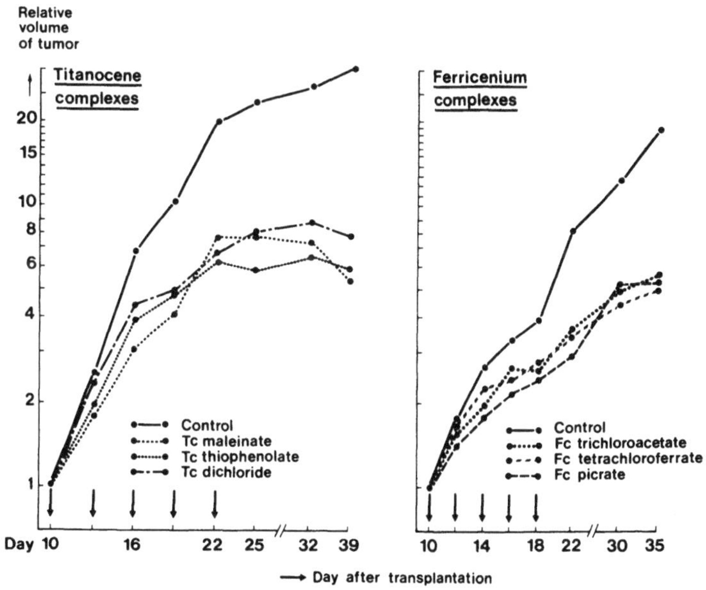

Fig. 2. Growth curves of a xenografted human lung adenocarcinoma after treatment with titanocene (Tc) and ferricenium (Fc) complexes. *(left)* Treatment with titanocene bis(hydrogenmaleinate) (Q3D × 5, D = 40mg/kg), titanocene bis(p-aminothiophenolate) bis(hydrochloride) (Q3D × 5, D = 60 mg/kg) and titanocene dichloride (Q3D × 5, D = 15mg/kg). *(right)* Treatment with ferricenium trichloroacetate mono(trichloroacetic acid)-solvated (Q2D × 5, D = 50mg/kg), ferricenium tetrachloroferrate (Q2D × 5, D = 70mg/kg) and ferricenium picrate (Q2D × 5, D = 80mg/kg)

Table 2. Growth inhibitions[a] of human xenografts by ferricenium complexes

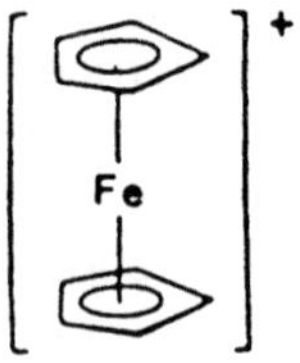 X^-	Colon adeno-carcinoma	Rectum adeno-carcinoma	Breast carcinoma	Lung adeno-carcinoma	Lung small cell carcinoma
Control	0	0	0	0	0
$X^- = [CCl_3COO]^- \cdot CCl_3COOH$	37	67	0	52	40
$X^- = [CCl_3COO]^- \cdot 2\ CCl_3COOH$	39	63	0	50	25
$X^- = \frac{1}{2}\ [Cl_3FeOFeCl_3]^{2-}$	64	75	54	30	51
$X^- = [FeCl_4]^-$	67	77	30	71	n.d.
$X^- = [2,4,6\text{-}(NO_2)_3C_6H_2O]^-$	58	83	n.d.	57	n.d.

[a] See explanation to Table 1.

Acknowledgement

This work was supported by a grant from MEDAC GmbH, Hamburg.

References

1. Koepf-Maier P, Hesse B, Koepf H: Tumorhemmung durch Metallocene: Wirkung von Titanocen-, Zirconocen- und Hafnocen-dichlorid gegenüber Ehrlich-Aszites-Tumor der Maus. J Cancer Res Clin Oncol 1980 (96): 43-51
2. Koepf-Maier P, Wagner W, Hesse B, Koepf H: Tumor inhibition by metallocenes. Activity against leukemias and detection of the systemic effect. Eur J Cancer 1981 (17): 665-669
3. Koepf-Maier P, Koepf H, Neuse EW: Ferricenium complexes: a new type of water-soluble antitumor agent. J Cancer Research Clin Oncol 1984 (109): 336-340
4. Koepf-Maier P: Tumor inhibition by ferricenium complexes: systemic effect in vivo and cell growth inhibition in vitro. Z Naturforsch 1985 (40c): 843-846
5. Koepf-Maier P, Preiss F, Marx T, Klapoetke T, Koepf H: Tumor inhibition by titanocene complexes: activity against sarcoma 180. Anticancer Res 1986 (6): 33-38
6. Koepf-Maier P, Koepf H: Tumor inhibition by titanocene complexes: activity against B16 melanoma and colon 38 carcinoma. Arzneimittelforsch (Drug Res) In press
7. Chang SI: Die Synthese von Metallverbindungen und experimentelle Untersuchungen über die tumorhemmende Wirkung von Substanzen aus der IV. Nebengruppe (translated from Chinese). Yao Hsueh T'ung Pao 1981 (16): 57
8. Koepf-Maier P: Unpublished results

VI. Human Tumour Xenografts in the Development of Biological Response Modifiers

The Effect of Human and Murine Interferons on Human Tumour Xenografts

F. R. Balkwill

Head Biological Therapy Laboratory, Imperial Cancer Research Fund, 44, Lincoln's Inn Fields, London WC2A 3PX, United Kingdom

Abstract

Human alpha interferons, HuIFN-alphas, caused tumour stasis and occasionally regression in a range of different human tumour xenografts growing subcutaneously in nude mice. The responses were seen at doses clinically achievable in man and were consistent for individual tumour lines over several passages. The HuIFN-alpha acted directly on the human tumour cells and had no effect on the murine host. HuIFN-gamma did not significantly affect the growth of subcutaneous xenografts but was capable of altering surface antigen expression in the xenografts. However, when xenografts were grown intraperitoneally or as lung tumour nodules, HuIFN-gamma had antitumour activity.

Murine IFN alpha/beta, but not murine IFN-gamma, could also cause stasis of subcutaneous xenografts but this effect was independent of the sensitivity of tumours to HuIFN alpha and appeared to be due to some modulation of host response to the tumour.

Human tumour necrosis factor, TNF, caused regression and cure of most subcutaneous xenografts when administered locally into the tumour. Intraperitoneal administration was however ineffective and the activity was not greatly enhanced when combined with HuIFN-alpha or gamma.

When TNF and HuIFN-gamma were combined to treat intraperitoneal xenografts of ovarian cancer synergistic interactions were seen and the majority of mice were cured of their tumour.

Thus the human tumour xenograft model can generate useful preclinical data on biological response modifiers.

Introduction

Biological response modifiers (BRMs) are defined as those agents or approaches that alter the relationship between the tumour and the host by modifying the host's biological response to the tumour cells with resultant therapeutic effects (1). It may seem inappropriate to study such agents in the human tumour xenograft model where the host immune system is defective and the tumour is of a different species, but there are reasons why such an approach can be successful. First, some BRMs, such as interferons (IFNs) and tumour necrosis factor (TNF) can have direct modulatory effects on tumour cells, and second, some BRMs, such as IFNs, are species specific and thus their effects on the host and the tumour can be dissociated in this model system. In this paper I will illustrate these points by reviewing the response of human tumour xenografts to IFNs alpha and gamma and TNF.

Interferons are the first in a series of BRMs that are undergoing preclinical and clinical evaluation for anticancer activity. First discovered as agents that will render a cell resistant to infection by a wide variety of agents, they are now conceived as cell regulators that play an important part in the control of immune responses and possibly normal growth and differentiation (2, 3).

Tumour necrosis factor was originally identified as an activity in the serum of endotoxin treated mice that would cause necrosis and cure of some experimental animal tumours. Like IFNs, TNF is now recognized as a cell regulator involved in immune responses; it has wide-ranging cell regulatory activity and is cytotoxic for many tumour cells (4).

Interferons and TNF form part of the lymphokine cascade and are produced during the same immune responses. A combination of the two is markedly synergistic *in vitro* (5, 6).

Effects of Interferons Alpha and Gamma on Human Tumour Xenografts in Nude Mice

Human interferons alpha (HuIFN-alpha) caused tumour stasis and occasionally tumour regression in a range of different human tumour xenografts derived

Table 1. Tumour growth inhibition by HuINFs - % of control tumour size index[a]

Tumour Code	Tumour Type	Hu-IFN alpha[b]	Hu-IFN gamma[c]
		(% of control)	
GFS	Bowel	70	119
GH6	Bowel	45	81
GFDe	Bowel	49	141
NcMc	Breast	17[d] (p = 0.005)	51
GFW	Bowel	2 (p = 0.002)	N.D.
GFC	Bowel	99	173
GFLa	Ovary	58 (p = 0.05)	75
EF3	Bowel	66	105
1068	Breast	40 (p = 0.05)	N.D.

[a] Tumour size index is the product of the 2 largest diameters at right angles.
[b] Hu-IFN-alpha (Wellferon) s. c.
[c] Recombinant HuIFN-gamma (Immuneron) i. p. given at 2×10^5 U/mouse/day daily for 28-40 days starting 2 weeks after tumour was implanted (see 11 for methods).
[d] statistically significant students T test.

All xenografts used before the 7th passage in nude mice.

from breast, bowel, lung, ovarian and renal carcinomas and osteosarcomas growing subcutaneously (s.c.) in nude mice (7-10). An example of typical responses is shown in Table 1. In this series of experiments responses ranged from complete regression of tumour and cure of the animal, through tumour stasis to progression at the same rate as control tumours. The reason for the differences between tumours is as yet unknown but was consistent for an individual tumour line over several passages. Responses were attained at doses equivalent to those clinically achievable in man (7).

The HuIFN-alpha appears to act directly on the human tumour cells and has no effect on the murine host (7, 11). This direct effect could be 1) on the growth of differentiation of the tumour cells, 2) on the response of the tumour cells to the environment, e.g. down regulation of growth factor receptors or 3) on the cell surface antigenicity of the tumour cells thus altering the host response to them.

It is of interest that replacement of osteosarcoma tissue with normal bone was seen in IFN-alpha treated xenografts (8).

Human IFN-gamma did not cause significant tumour stasis or regression in 7 different s.c. xenografts derived from primary breast, bowel and ovarian carcinomas, many of which were sensitive to HuIFN-alpha (Table 1). The reasons for this are not clear. It is possible that pure IFN-gamma is not cytostatic for as many cells as IFN-alpha (12), or it could be a problem of pharmacokinetics. In the nude mouse the pharmacokinetics of HuIFN-alpha and gamma are markedly different (11). However, we know that HuIFN-gamma can reach the tumour because it has at least one regulatory effect on some xenografts. The expression of Class II MHC antigens on the surface of ⅔ breast and bowel cancer xenografts was significantly induced or strongly enhanced by systemic therapy with HuIFN-gamma (13). This enhancement of MHC was exploited to promote localisation of a radiolabelled MHC Class II antibody to the tumour in HuIFN-gamma treated mice (14).

The enhancement of MHC Class II is achieved by lower doses of IFN-gamma than those generally required for cell growth inhibition, so it is still possible that insufficient IFN-gamma reached the subcutaneous site to actually inhibit tumour growth.

Certainly when xenografted material was grown in other sites in the nude mouse HuIFN-gamma, given intraperitoneally, was effective. Low doses of HuIFN-gamma inhibited the development of intraperitoneal xenografts of ovarian cancer (Balkwill & Ward, submitted for publication), and experimental pulmonary metastases of human melanoma cells (Ramani & Balkwill, submitted for publication).

Experiments using human IFNs and human tumour xenografts have therefore shown us that IFNs can have direct tumour cell growth inhibitory and cell surface modulatory effects, but because IFNs are species specific they have given us no information concerning IFNs modulation of host responses to the tumour. However, we have recently found that purified murine IFN alpha/beta, but not gamma (MuIFN), could cause significant cytostasis of human tumours growing subcutaneously (15). This response to MuIFN was independent of the sensitivity of tumours to HuIFN-alpha. Moreover, the tumour cells did not show any increase in host NK cell activity, was also seen in beige nude (NK deficient) mice, and was not accompanied by any host cell infiltration of the tumour (15). The mechanism of this host mediated effect on the human tumours is not yet understood but is probably not mediated by the host immune system. It could be the depletion of some vital host nutrient/growth factor required by the tumour, or some inhibition of the development of stromal support.

Effects of Tumour Necrosis Factor on Human Tumour Xenografts

Pure recombinant human TNF caused complete regression and cure of 1 breast and 2 bowel cancer xenografts when a low dose was injected intratumourally (i.t.). In a further 2 xenografts, cytostasis and regression occurred and only 1 xenograft was not sensitive. However, when the same dose was administered to the mice i.p., then only 2 of 6 tumours responded and the response was only cytostatic (16). In other experiments where an ovarian cancer cell line was grown in nude mice, intravenous (i.v.) administration of rHuTNF had some tumour static activity whereas i.p. therapy was ineffective (17). We have not yet compared i.v. to i.p. therapy.

Localised application of rHuTNF appears to be important in maximising its therapeutic activity in xenografted tumours. Intraperitoneal administration of rHuTNF did have significant antitumour activity against 2 of 4 ovarian cancer xenografts growing i.p. in nude mice (Balkwill & Ward, submitted for publication).

The mechanism(s) of the activity of rHuTNF are not yet understood. Unlike IFNs, TNF is not species specific (18) and rHuTNF would be expected to alter the activity of many different cells in the nude mouse host. When the regressing s.c. xenografts treated with i.t. rHuTNF were examined, a prominent infiltrate of host mononuclear and polymorphonuclear cells was seen (16). Similarly, when peritoneal washes from the mice bearing i.p. ovarian cancer xenografts were studied an increase in host polymorphs was seen (F. Balkwill, unpublished results).

Interactions Between Interferons and Tumour Necrosis Factor

Interferons, particularly INF-gamma, have been shown to synergise with TNF in cell culture assays (5). We have found some evidence for this synergy in the xenograft model. With primary tumours growing s.c. systemic TNF, inactive on its own, significantly improved the cytostatic activity of HuIFN-alpha in 3 of 3 xenografts studied, but only showed additive interactions with HuIFN-gamma in 1 of 3 xenografts studied (19). In 1 of the i.p. ovarian cancer xenografts studied dramatic synergy between HuIFN-gamma and rHuTNF was seen. When therapy was started 7 days after i.p. injection of the tumour cells, HuIFN-gamma or rHuTNF alone resulted in only 15 and 5%, respectively, cumulative survival at 155 days. A combination of the 2 resulted in 85% cumulative survival. All control mice died by 42 days (Balkwill & Ward, submitted for publication). Doses of rHuTNF or rHuIFN-gamma used were within the range achievable clinically (at least when used singly). In a further 3 different ovarian cancer xenografts the rHuTNF or rHuIFN-gamma were effective as single agents and further experiments are required to see if synergy can be demonstrated when very low doses are used.

Conclusions

Studies with the 2 different types of BRM, IFNs and TNF, have yielded some potentially useful preclinical information. Firstly, we know that IFNs can have direct effects on human tumours cells *in vivo*. Human IFN-alphas can be tumourstatic and in a minority of cases cause total regression of tumours. However, it is important to remember that in clinical trials a response is recorded when a tumour decreases in size by more than 50%. Such a response was only seen in a minority of xenograft experiments. Human IFN-gamma was able to change cell surface antigen expression in xenografts. This change may not be relevant to a xenograft but in a homologous host such a change on the tumour cell surface could promote host reactivity.

Secondly, we have highlighted differences between IFN-alpha and IFN-gamma. Unlike IFN-alpha, IFN-gamma had no antitumour activity on s.c. tumour xenografts but did act on xenografts growing i.p. or in the lungs. This finding indicates that careful consideration should be given to route of administration and pharmacokinetics of IFN-gamma. However, we have recently found that local i.t. administration of IFN-gamma to s.c. tumours did not increase its therapeutic activity (F. Balkwill, unpublished results).

Thirdly, we have evidence that an IFN-induced host-mediated, probably non-immune, mechanism can also cause tumour stasis of s.c. xenografts. The xenograft model has proved useful in dissecting direct and host mediated effects on IFNs on tumours.

Fourth, we have demonstrated that local administration of TNF is more effective, and finally we have demonstrated that in some xenografts efficacy of either group of BRMs can be improved by their combination.

Biological response modifiers have complex regulatory activities on both the tumour and host. Such activities can not be adequately studied *in vitro*. Human tumour xenografts are one animal model system which can be used to study the complexity of action of BRMs.

References

1. Oldham RK: Biological response modifiers. JNCI 1983 (70): 789-796
2. Balkwill FR: The regulatory role of interferons in the human response. In: Taylor-Papadimitriou J (ed) Interferons.

Their Impact in Biology and Medicine. New York: Oxford University Press 1985: 61-80
3. Resnitzky D, Yarden A, Zipori D, Kimchi A: Autocrine B-related interferons controls c-myc suppression and growth arrest during hematopoietic cell differentiation. Cell 1986 (46): 31-40
4. Beutler B, Cerami A: Cachectin and tumour necrosis factor as two sides of the same biological coin. Nature 1986 (320): 584-588
5. Fransen L, Van der Heyden J, Ruysschaert R, Fiers W: Recombinant tumour necrosis factor: its effect and its synergism with interferon-gamma on a variety of normal and transformed human cell lines. Eur J Cancer Clin Oncol 1986 (22): 419-426
6. Wong GHW, Goeddel DV: Tumour necrosis factor alpha and beta inhibit virus replication and synergize with interferons. Nature 1986 (323): 819-822
7. Balkwill FR: Antitumour effects of interferon in animals. In: Finter NB and Oldham RK (eds) Interferon: In vivo and clinical studies. Elsevier Science Publishers B.V. 1985
8. Brosjo O, Bauer HCF, Brostrom L, Nilsonne U, Nillsson OS, Reinholt FP, Strander H, Tribukait B: Influence of human alpha-interferon on four human osteosarcoma xenografts in nude mice. Cancer Res 1985 (45): 5598
9. Yamaoka T, Takada H, Ranagi Y, Kataoka T, Sakurai Y: The antitumour effects of human lymphoblastoid interferon on human renal cell carcinoma in athymic nude mice. Cancer Chemother Pharamacol 1985 (14): 184-187
10. Balkwill FR, Aldam G, Lee A: The action of interferons on human tumour xenografts growing in nude mice. In: Dianzani F, Rossi GB (eds) The Interferon System, Symposium Publications Raven Press 1986 (24): 327-331
11. Balkwill FR, Goldstein L, Stebbing N: Differential action of six human interferons against two human carcinomas growing in nude mice. Int J Cancer 1985 (35): 613-617
12. Saito T, Berens ME, Welander CE: Direct and indirect effects of human recombinant gamma-interferon on tumor cells in a clonogenic assay. Cancer Res 1986 (46): 1142-1147
13. Balkwill FR, Stevens MH, Griffin DB, Thomas JA, Bodmer JG: Interferon gamma regulates HLA-D expression on solid tumours in vivo. Eur J Cancer 1987 In press
14. Rowlinson G, Balkwill FR, Snook D, Gooker G, Epenetos AA: Enhancement by gamma-interferon of in vivo tumor radiolocalization by a monoclonal antibody against HLA-DR antigen. Cancer Res 1986 (46): 6413-6417
15. Balkwill FR, Proietti E: Effects of mouse interferon on human tumour xenografts in the nude mouse host. Int J Cancer 1986 (38): 375-380
16. Balkwill FR, Lee A, Aldam G, Moodier E, Thomas A, Tavernier J, Fiers W: Human tumour xenografts treated with recombinant human tumor necrosis factor alone or in combination with interferons. Cancer Res 1986 (46): 3990-3993
17. Creasey A, Reynolds MT, Laird W: Cures and partial regression of murine and human tumours by recombinant human tumour necrosis factor. Cancer Res 1986 (46): 5687-5690
18. Fransen L, Ruysschaert MR, Van der Heyden J, Fiers W: Recombinant tumor necrosis factor: species specificity for a variety of human and murine transformed cell lines. Cell Immunol 1986 (100): 260-267
19. Balkwill FR, Ward BG, Fiers W: Antitumour effect of TNF on human tumour xenografts in nude mice. In: Ciba symposium on tumour necrosis factor and related cytotoxins 1987 (In press)

Biodistribution and Therapeutic Efficacy of a Monoclonal Antibody-Methotrexate Conjugate in Mice

M.V. PIMM, J.A. CLEGG, M.C. GARNETT, and R.W. BALDWIN

Cancer Research Campaign Laboratories, University of Nottingham, Nottingam NG7 2RD, United Kingdom.

Abstract

A human tumour xenograft model has been used to assess the biodistribution and therapeutic potential of a methotrexate human serum albumin monoclonal antibody conjugate. The conjugate significantly retarded growth of a colon carcinoma xenograft whereas the same dose of free methotrexate was ineffective. Biodistribution studies with radiolabelled conjugate showed rapid clearance from plasma, particularly into the liver and spleen, suggesting that alternative conjugates with longer survival might show even greater therapeutic effects.

Introduction

It is now firmly established that a wide range of monoclonal antibodies directed against antigens preferentially associated with human tumours can localise *in vivo* in tumour deposits. This has been extensively investigated in nude mice with human tumour xenografts and these studies have formed the basis of clinical trials where tumour localization was initially demonstrated by external imaging of radiolabelled preparations and subsequently confirmed by analysis of radiolabel levels in resected tumour and normal tissues (reviewed in 1).

One of the major potential applications of tumour localising monoclonal antibodies is to preferentially deliver therapeutic agents to tumour sites. This may alter advantageously the therapeutic efficacy of existing cytotoxic agents and open the way to the use of newer toxic substances, particulary the plant and bacterial toxins. Already there are reports of therapy, albeit in experimental systems, with monoclonal antibodies conjugated to agents such as anthracyclines, (2), methotrexate (3, 4, 5), vindesine (6, 7), chlorambucil (8), and plant and bacterial toxins (10, 11, 12, 13).

In this Department conjugates of the murine monoclonal antibody 791T/36 have been produced with daunomycin (14), methotrexate (3, 4), vindesine (7), ricin toxin A chain (15) and interferon (16). These conjugates have been characterised for antibody directed toxicity against appropriate human tumour cells expressing the 791T/36 defined antigen. Human tumour xenograft models established in nude mice were initially used to evaluate tumour localization with monoclonal antibodies such as 791T/36 but also provide models to evaluate the therapeutic efficacy of these conjugates.

Although the basis of this approach is that the antibody will target drugs to tumour sites, the extravasation, biodistribution and catabolism of the antibody, already complex, might be altered significantly by drug conjugation. This needs to be appreciated in the design of drug-antibody conjugates to achieve maximum therapeutic usefulness. Results of studies on the biodistribution and therapeutic effect in nude mice of a methotrexate-human serum albumin-antibody (MTX-HSA-AB) conjugate will illustrate these problems.

Methods

The synthesis, structure and *in vitro* cytotoxicity of the MTX-HSA-AB conjugate have been extensively described elsewhere (3, 4). Briefly, it consists of a conjugate of 1 to 3 molecules of human serum albumin (HSA) per antibody molecule, in which each HSA molecule carries up to 30 methotrexate (MTX) molecules. Thus conjugate preparations are heterogeneous but overall have typical MTX:antibody molar rations of 30-40:1, and molecular weight of up to 340000 daltons. For *in vivo* studies, 791T/36 antibody or HSA alone or HSA-MTX conjugates were radiolabelled with ^{131}I or ^{125}I, and a MTX-HSA-AB conjugate was prepared in which the HSA moiety had been labelled with ^{125}I. Methotrexate in plasma was assayed using a commercial assay kit (Emit methotrexate assay, Syva Co., CA, U.S.A.).

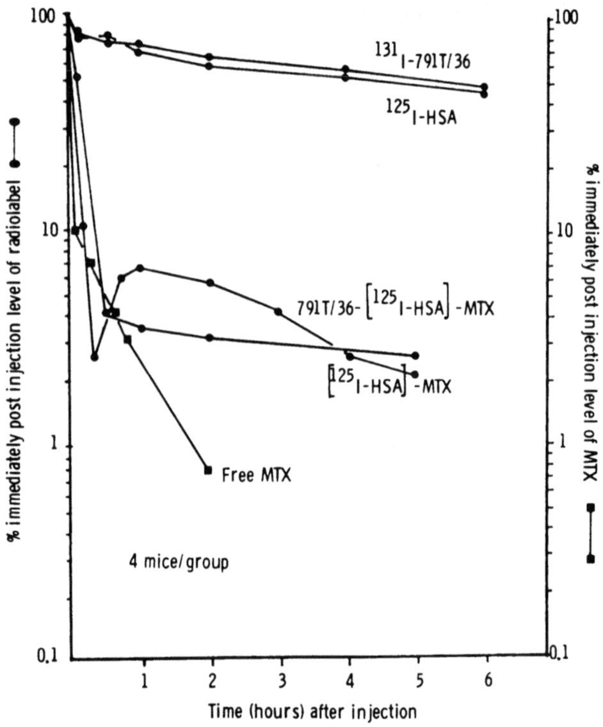

Fig. 1. Blood survival in mice of radiolabelled 791T/36 antibody (^{131}I-791T/36), HSA (^{125}I-HSA), HSA-MTX conjugate ([^{125}I-HSA-]-MTX) and MTX-AB conjugate (791T/36-[^{125}I-HSA]-MTX) as well as non protein-bound (free) MTX. Mice were injected intravenously with the preparations, and blood samples taken for analysis at regular intervals

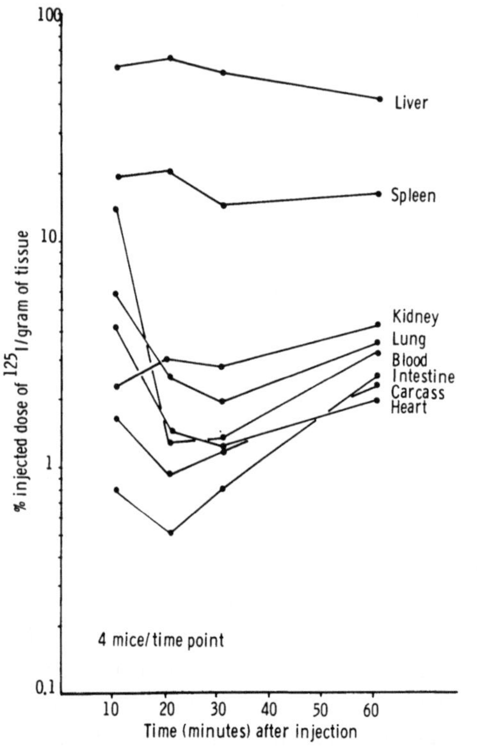

Fig. 2. Tissue levels of ^{125}I in mice following intravenous injection of methotrexate-[^{125}I-HSA-]-791T/36 antibody. Tissue levels shown as percent of injected dose/gram of tissue

Results

Figure 1 shows the plasma survival profiles in mice of radiolabelled antibody, HSA, HSA-MTX and MTX-HSA-AB conjugates up to six hours following intravenous injection. Although antibody and HSA showed prolonged survival, plasma levels of free methotrexate, HSA conjugated to methotrexate and MTX-HSA-AB conjugate fell very rapidly after injection. In some cases there was a transient rise in plasma levels of radiolabels, but this circulating radioiodine was not acid precipitable and probably represents the presence of free radioiodine released following conjugate catabolism.

Analysis of organ levels of radiolabel showed that the liver and spleen were major organs of clearance of the MTX-HSA-AB conjugate, levels here exceeding those of blood within minutes of injection (Fig. 2). In mice with xenografts of colon carcinoma C170, in which tumour localisation of radiolabelled antibody alone has previously been well characterised (17), there was no tumour localization of radiolabel within one to four days of injection of MTX-HSA-AB conjugate in which the HSA was radiolabelled (Fig. 3), these being optimum times for tumour localization of antibody alone.

When tissue levels were normalized with respect to the whole body retention of radiolabels, three days after injection, blood levels of both radiolabelled antibody and HSA exceeded those of other organs. After the administration of HSA-MTX and the full MTX-HSA-AB conjugate blood levels were no

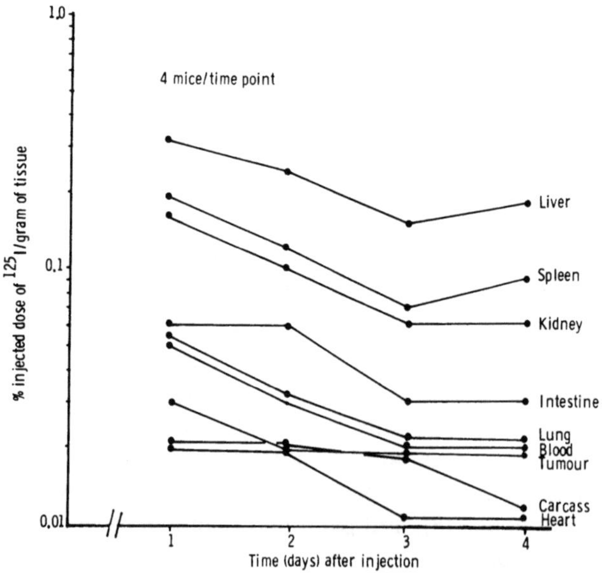

Fig. 3. Tissue and tumour levels of ^{125}I in nude mice with xenografts of colon carcinoma C170 following intravenous injection of methotrexate-[^{125}I-HSA]-791T/36 antibody. Tissue levels shown as percent injected dose/gram of tissue

higher than most organs, and markedly lower than those of spleen, liver and kidney (Fig. 4).

Therapeutic assessment of MTX-HSA-791T/36 conjugates have shown effects against xenografts of human osteosarcoma and colon carcinoma expressing the 791T/36 defined antigen (1). Growth of colon carcinoma C170 was significantly retarded following administration of conjugate, while free methotrexate at equivalent doses was ineffective (Fig. 5).

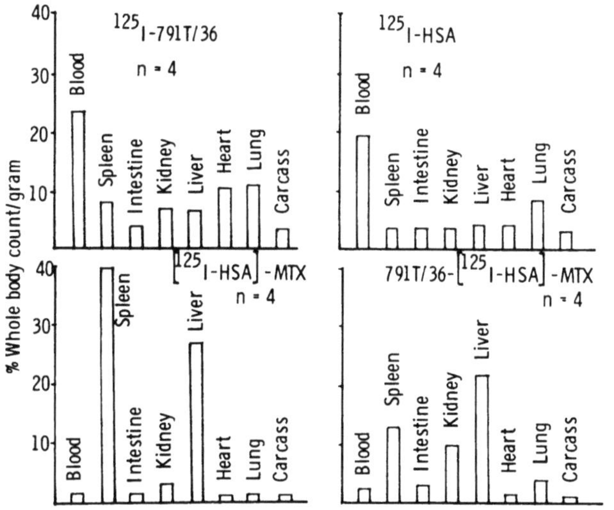

Fig. 4. Tissue levels of ^{125}I in mice following intravenous injection of ^{125}I labelled preparations three days previously. Tissue levels shown as the proportion of the whole body ^{125}I count rate/gram of tissue

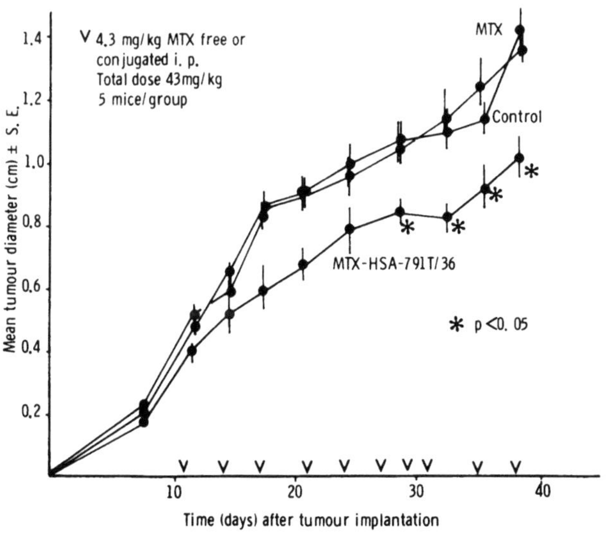

Fig. 5. Therapeutic effect against xenografts in nude mice with colon carcinoma C170. Only conjugated drug (MTX-HSA-AB) showed significant retardation of growth

Discussion

The above described and related studies (18) have shown that, while the methotrexate-HSA-791T/36 conjugate can exert therapeutic effects, the biodistribution and catabolism is unlike that of free antibody or HSA. This suggests that therapeutic effects which are being achieved result from tumour localisation of only minute amounts of conjugated drug. For conjugated drug, either HSA-MTX or MTX-HSA-791T/36, clearance from the blood stream is rapid, most of the uptake occurring in liver and spleen. Methotrexate is also taken up by the liver (19), probably mediated by hepatocytes and by renal tubular cells. This uptake can be reduced by the administration of probenecid which inhibits net cellular methotrexate influx and thus hepatic and renal clearance of methotrexate (20, 21). Conversely, ricin is cleared primarily by hepatic Kuppfer cells via mannose receptors since this toxin contains a significant level of mannose (22). Deglycosylation of ricin (11, 12) or host blockade with mannan or ovalbumin which contains similar residues (22) have both been reported to reduce hepatic clearance and enhance therapeutic effects of conjugated ricin or ricin A chain. The site of clearance of conjugated methotrexate is not yet known, but clearly there is great scope for modification of conjugates to reduce host clearance mechanisms. Such modifications should result in a more favourable biodistribution and increased therapeutic effects using similar treatments.

These studies were supported by the Cancer Research Campaign U.K.

References

1. Baldwin RW and Byers VS (eds): Monoclonal antibodies for cancer detection and therapy. London: Academic Press 1985
2. Pimm MV, Jones JA, Price MR, Middle JG, Embleton MJ, Baldwin RW: Tumour localization of monoclonal antibody against a rat mammary carcinoma and suppression of tumour growth with adriamycin-antibody conjugates. Cancer Immunol Immunother 1982 (12): 125-134
3. Garnett MC, Embleton MJ, Jacobs E and Baldwin RW: Preparation and properties of a drug-carrier-antibody conjugate showing selective antibody-directed cytotoxicity in vitro. Int J Cancer 1983 (31): 661-670
4. Garnett MC, Embleton MJ, Jacobs E and Baldwin RW: Studies on the mechanism of action of an antibody-targeted drug-carrier conjugate. Anti-Cancer Drug Design 1985 (1): 3-12
5. Kulkarni P, Blair AH, Ghose T, Mammen M: Conjugation of methotrexate to IgG antibodies and their F(ab)2 fragments and the effect of conjugated methotrexate on tumour growth in vivo. Cancer Immunol Immunother 1985 (19): 211-216

6. Kanellos J, Pietersz GA, McKenzie IFC: Studies of methotrexate-monoclonal antibody conjugates for immunotherapy. J Natl Cancer Inst 1985 (75): 319-332
7. Rowland GF, Axton CA, Baldwin RW, Brown JR, Corvalan JRF, Embleton MJ, Gore VA, Hellstrom I, Hellstrom KE, Jacobs E, Marsden CH, Pimm MV, Simmonds RG and Smith W: Antitumour properties of vindesine-monoclonal antibody conjugates. Cancer Immunol Immunother 1985 (19): 1
8. Rowland GF, Simmonds RG, Gore VA, Marsden CH and Smith W: Drug localization and growth inhibition studies of vindesine-monoclonal anti-CEA conjugates in a human tumour xenograft. Cancer Immunol Immunother 1986 (21): 183-187
9. Smyth MJ, Pietersz GA, Classon BJ, McKenzie IFC: Specific targeting of chlorambucil to tumours with the use of monoclonal antibodies. J Natl Cancer Inst 1986 (76): 503-510
10. Blythman HE, Casellas P, Gros O, Gros P, Jansen FK, Paolucci F, Pau B, Vidal H: Immunotoxins: Hybrid molecules of monoclonal antibodies and a toxin subunit specifically kill tumour cells. Nature 1981 (290): 145-146
11. Thorpe PE, Brown ANF, Bremner JAG, Foxwell BMJ, Stirpe F: An immunotoxin composed of monoclonal antithy 1.1 antibody and a ribosome-inactivating protein from Saponaria officinalis: potent antitumour effects in vitro and in vivo. J Natl Cancer Inst 1985 (75): 151-159
12. Thorpe PE, Detre SI, Foxwell BMJ, Brown ANF, Skilleter DN, Wilson G, Forrester IA and Stirpe F: Modification of the carbohydrate in ricin with metaperiodate-cyanborohydride mixtures. Eur J Biochem 1985 (147): 197-206
13. Weil-Hillman G, Runge W, Jansen FK, Vallera DA: Cytotoxic effect of antibo-M 67,000 protein immunotoxins on human tumours in a nude mouse model. Cancer Res 1985 (45): 1328-1336
14. Gallego J, Price MR, Baldwin RW: Preparation of four daunomycin antibody 791T/36 conjugates with anti-tumour activity. Int J Cancer 1984 (33): 737
15. Embleton MJ, Byers VS, Lee HM, Scannon P, Blackhall N and Baldwin RW: Sensitivity and selectivity of ricin toxin A chain-monoclonal antibody 791T/36 conjugated against human tumor cell lines. Cancer Res 1986 (46): 5524
16. Pelham JM, Gray JD, Flannery GR, Pimm MV, Baldwin RW: Interferon-alpha conjugation to human osteogenic sarcoma monoclonal antibody 791T/36. Cancer Immunol Immunother 1983 (15): 210-216
17. Durrant LG, Robins RA, Pimm MV, Perkins AC, Armitage NC, Hardcastle JD and Baldwin RW: Antigenicity of newly established colorectal carcinoma cell lines. Br J Cancer 1986 (53): 37-46
18. Baldwin RW, Embleton MJ, Gallego J, Garnett M, Pimm MV and Price MR: Monoclonal antibody drug conjugates for cancer therapy. In: Roth J (ed) Monoclonal Antibodies in Cancer, Advances in Diagnosis and Treatment. Futura Publishing Co. 1986
19. Kates RE, Tozer TN and Sorby DL: Increased methotrexate toxicity due to concurrent probenecid administration. Biochem Pharmacol 1976 (25): 1485-1488
20. Gewirtz DA, Plotkin JH and Randolph JC: Interaction of probenecid with methotrexate transport and release in the isolated rat hepatocyte in suspension. Cancer Res 1984 (44): 3846-3850
21. Sirotnak FM, Moccio DM, Hancock CH and Young CW: Improved methotrexate therapy of murine tumors obtained by probenecid-mediated pharmacological modulation at the level of membrane transport. Cancer Res 1981 (41): 3944-3949
22. Bourrie BJP, Casellas P, Blythman HE, Jansen FK: Study of the plasma clearance of antibody-ricin-chain immunotoxins. Eur J Biochem 1986 (155): 1-10

The Effect of Interferon-Alpha on the Activity of Cytotoxic Drugs in Bronchial Carcinoma Xenografts

R. J. FERGUSSON and J. F. SMYTH

Imperial Cancer Research Fund, Medical Oncology Unit, Department of Clinical Oncology, Western General Hospital, Edinburgh EH4 2XU, Scotland, United Kingdom

Abstract

Two human non-small cell lung cancer xenograft lines were used to study the activity of combinations of 2 cytotoxic drugs with human alpha-2 interferon. Statistically significant potentiation of cisplatin and ifosfamide given weekly for 5 weeks in a low dose was seen when recombinant human alpha interferon (rHu IFN-alpha 2b) was administered simultaneously (2×10^4 U/mouse/day). Median doubling time for cisplatin treated tumours increased from 17 and 14 days to 27 and 18.5 days with the addition of interferon, whereas for ifosfamide they increased from 11.5 and 14 days to 15 and 16 days. Human alpha-2 interferon at this dose had no effect as a single agent.

Introduction

The outlook for the majority of patients with inoperable non-small cell lung cancer remains poor (1). Improved response rates and survival have been reported recently with regimes containing cisplatin (CDDP) (2), although the considerable toxicity associated with these combinations may preclude their widespread use (3, 4). Human interferons (IFNs) have been shown to have modest anti-tumour activity in a number of malignancies (5-8), although no activity has been reported in non-small cell lung cancer (NSCLC) (9-12).

Recently much interest has been shown in combining IFNs with standard chemotherapeutic agents. Several workers have shown potentiation of cytotoxic drugs by IFNs *in vitro* (13, 14), in animal tumour systems (15, 16) and in human tumours grown in immunodeficient mice (17). The aim of this study was to assess the effect of IFN-alpha on activity of 2 cytotoxic drugs in the treatment of 2 human non-small cell lung cancer xenografts growing in immune deprived mice.

Materials and Methods

Animals

CBA/lac mice were thymectomised at 3 weeks of age and then underwent whole body x-irradiation (7.35 Gy via a 250 KV, 15 mA source at 0.37 Gy/min +/− 2%) 3 weeks later. Forty-eight hours before irradiation a priming dose of cytosine arabinoside (200 mg/kg i.p.) was administered to protect the bone marrow and gastrointestinal tract from lethal radiation toxicity (18). The mice were housed in a separate room in the animal unit but specific pathogen-free conditions were not required.

Tumours

Two non-small cell lung cancer xenografts were used: (a) NX002, a moderately differentiated squamous carcinoma; (b) CX117, a poorly differentiated adenocarcinoma. NX002 was obtained from a bronchial biopsy and CX117 from a thoracotomy in previously untreated patients. Their growth characteristics and morphology have been described elsewhere (19). The tumours were tested during the 5th-11th passages. Both maintained the histological features of their original tumour and the karyotypes were confirmed to be human.

Xenografts were measured twice weekly using calipers, and tumour volume was estimated assuming an ellipsoid shape. Only tumours attaining a volume of 0.3 cm^3 were used for drug testing. A minimum of 6 tumours were entered in each treatment group.

Interferon

Human recombinant alpha-2 interferon (rHuIFN-α2b) was kindly provided by the Schering Corporation (Bloomfield, NJ, USA). It had a specific activity of 1.7×10^8 units/mg and was more than 98% pure. Doses of 2×10^4 units in 0.1 ml were injected s.c.

each day for 35 days at a site distant to the tumours. This dose is equivalent to approximately 2×10^6 units/m²/day in humans (20).

Cytotoxic Drugs

Cisplatin (Bristol-Myers Ltd, Slough, UK) was dissolved in sterile normal saline and administered at a dose of 1.4 mg/kg i.p. on days 0, 7, 14, 21 and 28. This dose was equivalent to 20% of the maximum tolerated dose previously determined in these animals. Ifosfamide (IFOS) (Boehringer Ingelheim, Bracknell, UK) was dissolved in sterile water for injection and administered at a dose of 60 mg/kg i.p. following an identical schedule.

Study Design

Groups of tumours were randomly allocated to treatment with drug or IFN alone, drug and IFN in combination, or to a control group. Response to treatment was assessed by measurement of the median doubling time (T_D) for each group and calculation of the specific growth delay (SGD) using the formula (21):

$$SGD = \frac{T_D(\text{treated}) - T_D(\text{control})}{T_D\ (\text{control})}$$

The SGD represents the number of tumour doubling times delayed by treatment and allows comparisons to be made between drugs in tumours of different growth rates. Statistical differences between the mean T_Ds of treated and control groups were carried out using an unpaired student's t test.

Results

The effect of rIFN-alpha 2b and the 2 drugs, either alone or in combination, on the median T_D and SGD in the 2 tumours is illustrated in Table 1 and Figures 1 and 2.

Table 1. Antitumour effect on the 2 xenografts studied

Drug	Dose (units)	(mg/kg)	Schedule	NX002 T_D[a]	NX002 SGD[a]	CX117 T_D	CX117 SGD
Control	-	-	-	9	-	12.5	-
rIFN	2×10^4	-	d × 35	10	0.11	10	0.00
CDDP	-	1.4	q7d × 5	17	0.88	14	0.12
CDDP + rIFN	2×10^4	1.4	q7d × 5 d × 35	27	2.0	18.5	0.48
IFOS	-	60	q7d × 5	11.5	0.28	14	0.12
IFOS + rIFN	2×10^4	60	q7d × 5 d × 35	15	0.67	16	0.28

[a] Median values of at least 6 tumours/group.

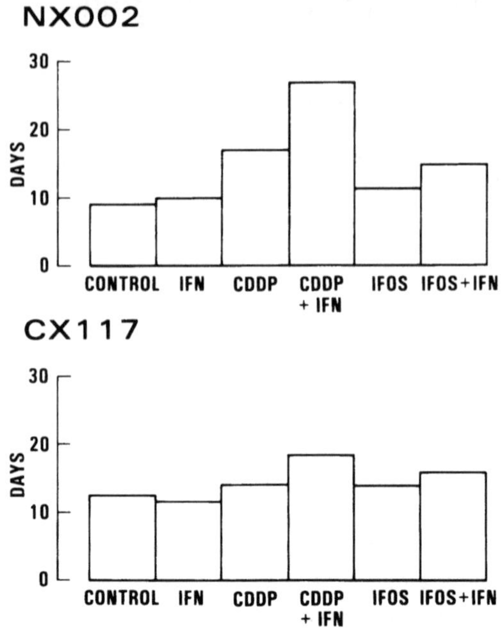

Fig. 1. Effect of rIFN-alpha 2b, CDDP, IFOS and rIFN-alpha 2b/CDDP and IFOS combinations on the median doubling time of the 2 xenografts studied. Minimum of 6 tumours/group

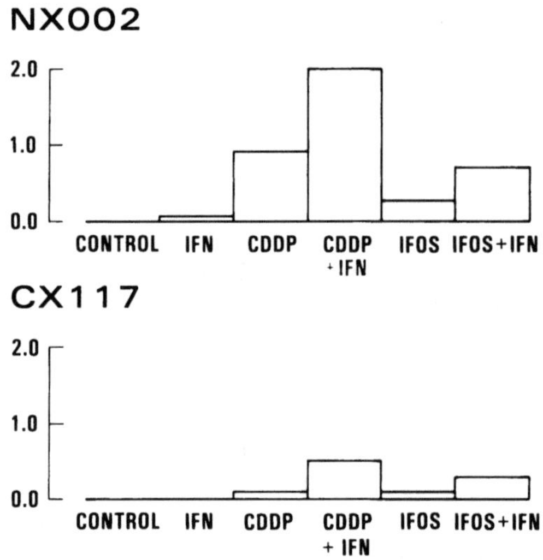

Fig. 2. Effect of rIFN-alpha 2b, CDDP, IFOS and rIFN-alpha 2b/CDDP and IFOS combinations on the specific growth delay of the 2 xenografts studied. Minimum of 6 tumours/group

NX002

The effect of rIFN-alpha 2b, CDDP and IFOS on tumour growth is shown in Table 1 and Figures 3 and 4. The median T_D for control tumours was 9 days. rIFN-alpha 2b had no effect as a single agent. Treatment with CDDP alone caused a significant increase in mean T_D compared with controls ($p < 0.04$). The addition of rIFN-alpha 2b to CDDP in-

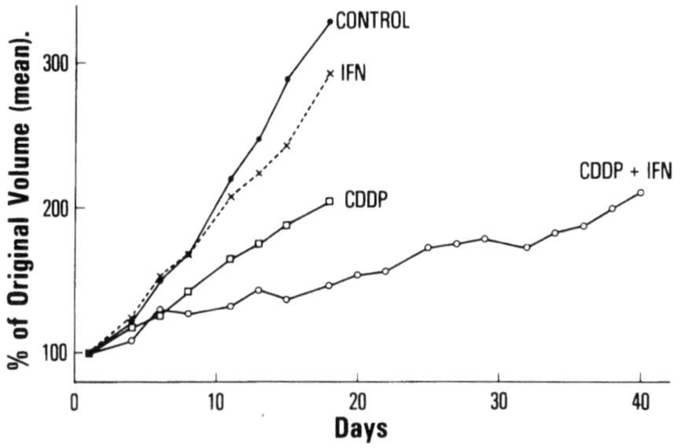

Fig. 3. Inhibitory effects of rIFN-alpha 2b/CDDP combination, compared with single agent activity on the growth of the squamous lung cancer xenograft NX002. Results are up until time of tumour doubling

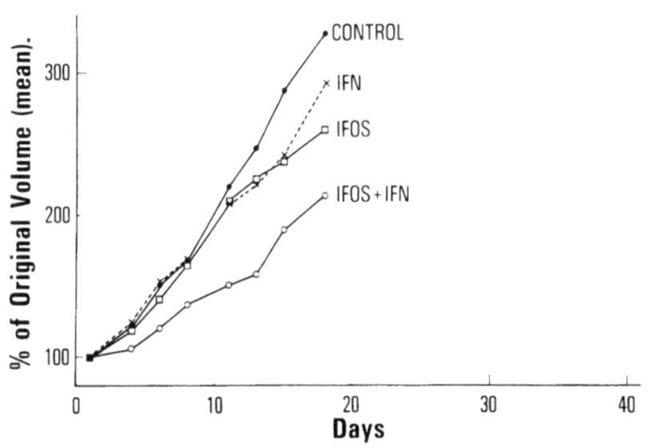

Fig. 4. Inhibitory effects of rIFN-alpha 2b/IFOS combinations, compared with single agent activity, on the growth of NX002. Results are up until time of tumour doubling

creased the potency further with a more significant prolongation of mean T_D ($p<0.005$). Three tumours (1 CDDP alone, 2 CDDP/rIFN-alpha 2b) regressed completely.

IFOS had little activity alone but when rIFN-alpha 2b was added an increased effect was seen. The mean T_D for the rIFN-alpha 2b/IFOS combination was significantly prolonged compared with controls ($p<0.04$) and compared with IFOS alone ($p<0.05$).

CX117

Median T_D for control tumours was 12.5 days. rIFN-alpha 2b had no effect as a single agent. CDDP and IFOS alone caused small increases in median T_D to 14 days. CDDP and rIFN-alpha 2b in combination gave a significant prolongation of mean T_D compared with controls ($p<0.02$). The addition of rIFN-alpha 2b to IFOS did not significantly improve its potency.

Discussion

In 2 human NSCLC xenografts we have demonstrated potentiation of chemotherapy with 2 drugs by the addition of recombinant human alpha-2 interferon. Injected in a dose of 2×10^4 units/mouse/day, rIFN-alpha 2b had no effect alone but was able to augment the activity of CDDP in CX117 and of CDDP and IFOS in NX002. The equivalent dose of IFN in humans (2×10^6 units/day) could be tolerated by most patients. These experiments confirm previous reports of enhanced activity seen with combinations of chemotherapy and IFNs in certain murine tumours (15, 16) and in a human breast cancer xenograft (17). To our knowledge this effect has not been reported in human lung cancer.

Although we have demonstrated that the combination of rIFN-alpha 2b and chemotherapy is more effective than would be expected from the potency of each as a single agent, it is not possible to prove a synergistic relationship from these data. This requires performing full dose/response experiments for each agent and combination of agents. However, the fact that rIFN-alpha 2b had no effect alone yet was able to increase the activity of both CDDP and IFOS in combination, is strong evidence of a positive or synergistic interaction.

The use of a xenograft system for drug testing has been validated by a close correlation between experimental results of the maximum tolerated dose (MTD) in mice and the therapeutic activity of the same cytotoxic drugs in the original tumours in patients (22). In the experiments reported here we divided the MTD into 5 aliquots given at weekly intervals in order to assess the response over a relatively chronic exposure with IFN. Further studies are now in progress to define the optimal dose schedule for such combinations. In clinical practice CDDP causes significant toxicity on its own (2, 3) and in clinical trials caution will have to be exercised to avoid potentiation of unacceptable toxicity resulting from CDDP-IFN combinations.

Since the validity of the xenograft model depends on the immune deprivation of the mouse host, we are aware of the possible immunostimulant effects of IFN that might artificially appear to potentiate the cytotoxicity of anticancer drugs in this system. It is known that murine IFN affects natural killer cell activity (23) and depresses the activity of the hepatic drug metabolising enzymes in the mouse (24). These effects have to be considered when murine IFN is tested in combination with drugs in experimental murine cancers (15, 16). Such effects are not seen when human IFN and anticancer agents are used to treat human tumours grown in immunodeficient mice (25). This suggests that the interactions between IFN and chemotherapy in such a system are likely to be due to a direct action on the tumour itself rather than due to an influence on the host tissues. However, the precise mechanism of IFN potentiation of chemotherapy remains unclear and is the subject of studies currently in progress.

References

1. Souhami RL: The management of advanced non-small cell carcinoma of the bronchus. In: Smyth JF (ed) The management of lung cancer. London: Edward Arnold 1984: 132-149
2. Gralla RJ, Casper ES, Kelsen DP, Braun DW, Dukeman ME, Martini N, Young CW and Golbey RB: Cisplatin and vindesine combination chemotherapy for advanced carcinoma of the lung: a randomized trial investigating two dosage schedules. Ann Intern Med 1981 (95): 414-420
3. Carmichael J, Gregor A, Cornbleet MA, Allan SG, McIntyre MA, Grant IWB, Crompton GK, Leonard RCF and Smyth JF: Vindesine and cis-platinum in the treatment of non-small cell lung cancer. Eur J Cancer Clin Oncol 1985 (21): 811-814
4. Souhami RL: Chemotherapy in non-small cell bronchial carcinoma. Thorax 1985 (40): 641-645
5. Quesada JR, Swanson DA, Trinidade A and Gutterman JU: Renal cell carcinoma: anti-tumour effects of leukocyte interferon. Cancer Res 1983 (43): 940-947
6. Merigan TC, Sikora C, Breeden JH, Levy R and Rosenberg SA: Preliminary observations of the effect of human leukocyte interferon in non-Hodgkin's lymphoma. N Engl J Med 1978 (299): 1449-1453
7. Quesada JR, Reuben J, Manning JT, Hersh EM and Gutterman JU: Alpha interferon for induction of remission of hairy-cell leukemia. N Engl J Med 1984 (310): 15-18
8. Real FX, Krown SE, Krim M, Myskowski PL, Safai B and Oettgen HR: Treatment of Kaposi's sarcoma with recombinant leukocyte A interferon. Proc Am Soc Clin Oncol 1984 (3): 55
9. Grunberg SM, Kempf RA, Itri LM, Venturi CL, Boswell WD and Mitchell MS: Phase II study of recombinant alpha interferon in the treatment of advanced non-small cell lung carcinoma. Cancer Treat Rep 1985 (69): 1031-1032
10. Figlin RA and Sarna GP: Human leukocyte interferon: phase II trial in non-small cell lung cancer and adenocarcinoma of the colon/rectum. Proc Am Soc Clin Oncol 1983 (2): 45
11. Leavitt RD, Duffey P and Aisner J: A phase II study of recombinant leukocyte-A interferon in non-small cell carcinoma of the lung. Proc Am Soc Clin Oncol 1984 (3): 52
12. Ernst P, Olesen BK, Nissen MH, Hokland P, Justesen J and Hansen HH: In vivo effect of recombinant interferon alpha in patients with lung cancer. Proc Am Soc Clin Oncol 1984 (3): 66
13. Aapro MS, Alberts DS and Salmon SE: Interactions of human leukocyte interferon with vinca alkaloids and other chemotherapeutic agents against human tumours in a clonogenic assay. Cancer Chemother Pharmacol 1983 (10): 161-166
14. Welander CE, Morgan TM, Homelsey HD, Trotta PP and Spiegel RJ: Combined recombinant human interferon alpha-2 and cytotoxic agents studied in a clonogenic assay. Int J Cancer 1985 (35): 721-729
15. Chirigos MA and Pearson JW: Cure of murine leukemia with drug and interferon treatment. J Natl Cancer Inst 1973 (51): 1367-1368
16. Gresser I, Maury C and Tovey M: Efficacy of combined interferon-cyclophosphamide therapy after diagnosis of lymphoma in AKR mice. Eur J Cancer 1978 (14): 97-99
17. Balkwill FR and Moodie EM: Positive interactions between human interferon and cyclophosphamide or adriamycin in a human tumor model system. Cancer Res 1984 (44): 904-908
18. Steel GG, Courtenay VD and Rostom AY: Improved immunosuppression techniques for the xenografting of human tumours. Br J Cancer 1978 (37): 224-230
19. Fergusson RJ, Carmichael J and Smyth JF: Human tumour xenografts growing in immunodeficient mice: a useful model for assessing chemotherapeutic agents in bronchial carcinoma. Thorax 1986 (41): 376-380
20. Freireich EJ, Gehan RA, Rall DA, Schmidt LH and Skipper HE: Quantitative comparison of toxicity of anticancer agents in mouse, rat, hamster, dog, monkey and man. Cancer Chemother Rep 1966 (50): 219-244
21. Kopper L and Steel GG: The therapeutic response of three human tumor lines maintained in immune-suppressed mice. Cancer Res 1975 (35): 2704-2713
22. Shorthouse AJ, Peckham MJ, Smyth JR and Steel GG: The therapeutic response of bronchial carcinoma xenografts: a direct patient-xenograft comparison. Br J Cancer 1980 (41): 142-145
23. Djeu JY, Heinbaugh JA, Holden HT and Heberman RB: Augmentation of mouse natural killer cell activity by interferon and interferon inducers. J Immunol 1979 (122): 175-181
24. Singh G, Renton KW and Stebbing N: Homogenous interferon from E. Coli depresses hepatic cytochrome P-450 and drug biotransformation. Biochem Biophys Res Commun 1982 (106): 1256-1261
25. Balkwill FR, Moodie EM, Freedman V and Fantes KH: Human interferon inhibits the growth of established human breast tumours in nude mice. Int J Cancer 1982 (30): 231-235

The Nude Mouse for the Study of Immunotoxins

H. E. BLYTHMAN, A. BORD, I. BUISSON, O. THURNEYSSEN, G. RICHER, and F. K. JANSEN

Centre de Recherches Clin-Midy/SANOFI, 34082 Montpellier Cedex, France

Introduction

Our laboratory is involved in the study of immunotoxins (IT), hybrid molecules made of 2 moieties: a toxic fragment of the plant toxin ricin, coupled by chemical means to a carrier, a monoclonal antibody.

When the antibody chosen is directed against antigenic specificities found on human tumour cells, *in vivo* studies must be performed using the adequate xenografts with the target cells for the immunotoxin. Setting up *in vivo* models poses problems for several tumour types. Some cell lines or explants require additional irradiation of nude mice before cell transfer, e.g. T-cell leukaemias.

Even though the direct extrapolation from the animal model to the clinical situation is quite fallacious, the use of the nude mouse can provide some helpful information on the efficiency of IT, the pharmacokinetics of the molecule and its constituents.

Table 1. Attempts to set up the CEM model; nu/nu BALB/c mice from Charles Rivers France were used

Cell dose	SC	IV	Xrays[a]	Takes
$50 \cdot 10^6$	+	ND	–	6/6
			300 R	5/6
			500 R	6/6
$30 \cdot 10^6$	+	ND	–	1/6
$20 \cdot 10^6$	+	ND	–	0/6
$10 \cdot 10^6$	+	ND	–	3/6
			300 R	6/6
			500 R	6/6
$5 \cdot 10^6$	+		500 R	0/5
		+	500 R	4/10
$2.5 \cdot 10^6$	+	ND	500 R	0/5
		+	400 R D-3	0/6
$1 \cdot 10^6$	ND	+	500 R	3/5
$5 \cdot 10^5$	ND	+	500 R	0/10
			400 R D-3	0/6

[a] Xrays given 24 hrs prior to cell injection.

Material and Methods

"Nude" mice (nu/nu BALB/c) from 2 breeders were used: Charles Rivers France (from their USA breeding colony) and Gl. Bomholtgaard Ltd (now BOM-MICE), Ry, Denmark.

Several established human T-cell leukaemia cell lines were tested to set up experimental models bearing the T65 (T1 or CD5) antigen: Jurkat, JM, HPB-ALL, RPMI-8402, CEM and P12-ICHIKAWA. Cells were maintained in continuous culture, except for P12-ICHIKAWA, which must be maintained by regular passage in nude mice, as ascites (1).

The immunotoxin tested in the animal models was obtained with the monoclonal antibody T101 (supplied by Hybritech) (2), coupled to the toxic moiety of the plant toxin ricin (3). T101 antibody recognizes the CD5 antigen present on the cell lines used.

Table 1 summarizes our attempts to set up a reproducible model with CEM cells injected into Charles Rivers nude mice: several cell doses were used, and different irradiation doses were also employed. The results were deceiving, e.g., giving non-reproducible and non-dose related takes, therefore the model was abandoned.

The best model in our hands was obtained with P12-ICHIKAWA cells, injected i.p. or i.v. This cell line was originally established from the peripheral blood of a 7-year old patient with ALL (1). The advantage of this cell line is that good and reproducible results can be obtained without irradiating the animals before cell injection, even though higher cell numbers are needed to induce the disease. Two different administration routes were used: i.p. and i.v. I.p. inoculation induces more or less haemorrhagic ascites, whereas i.v. inoculation gives rise to 2 variants of the disease; one with neurological involvement, presenting with hind leg paralysis, the other with the appearance of subcutaneous masses at the level of the neck and between the shoulder blades. Therefore, this model was chosen for our studies on immunotoxins.

Results

The P12-ICHIKAWA Model - Survival Curves

Figure 1 presents cell dose dependent survival curves obtained with these cells injected into nude mice. Male and female nu/nu BALB/c BOM were used; cells were injected i.p. or i.v., and mortality was recorded.

Treatment of the P12-ICHIKAWA Disease by Specific Immunotoxin

With this model, it has been possible to study the therapeutic potential of the T101 immunotoxin. Figure 2 presents preliminary experiments, with cells injected either i.p. or i.v. In the first experiment (Fig. 2a) 25×10^6 cells were injected i.p. on day zero. Specific immunotoxin (T101) was injected, also i.p., on the 8 following days, for a total dose of 12 µg of bound A-chain. Control animals received either PBS or a mixture of free A-chain and free specific antibody T101, in the same amounts as the specific immunotoxin (15 mice per group). Even if no survival was obtained, a significant delay of mortality was observed in the immunotoxin-treated group.
In another experiment (Fig. 2b), 10×10^6 cells were injected i.v. on day zero. Treatment was given i.p. for the 4 following days, for a total dose of 400 µg of bound A-chain. Controls received either PBS, free antibody (T101) alone or a nonspecific (antimelanoma) immunotoxin. While nonspecific immunotoxin gave results comparable to the PBS group, a slight beneficial effect was observed with free antibody. The best survival was obtained with the specific immunotoxin, with 50% mortality recorded on day 196, whereas control groups presented with 50% mortality between days 53 and 71. At the end of the experiment (220 days), 4 mice were alive and well.
This model has also enabled the study of the possibility to increase the efficiency of the immunotoxin, by the use of "potentiating" agents such as the carboxylic ionophore Monensin. Monensin can be conjugated to a carrier protein such as human serum albumin (HSA) (4) which renders the monensin molecule water-soluble, therefore of easy use. Figure 3 presents the results of experiments designed to test this possibility.
In the first part of the figure (3a), 10×10^6 cells were injected i.v. on day zero. Treatment was also i.v., in a single dose of 200 µg bound A-chain, given on day 1. Controls were the same as for Figure 2b (PBS, free A-chain alone at the same dose or nonspecific immunotoxin). This protocol did not induce a significant therapeutical effect, in contrast to what was obtained when the same total amount of immunotoxin was given i.p., over 4 consecutive days (Fig. 2b).
Fig. 3b shows the results obtained when the potentiating compound HSA-SS-Monensin was added to the immunotoxin treatment. Ten million cells were

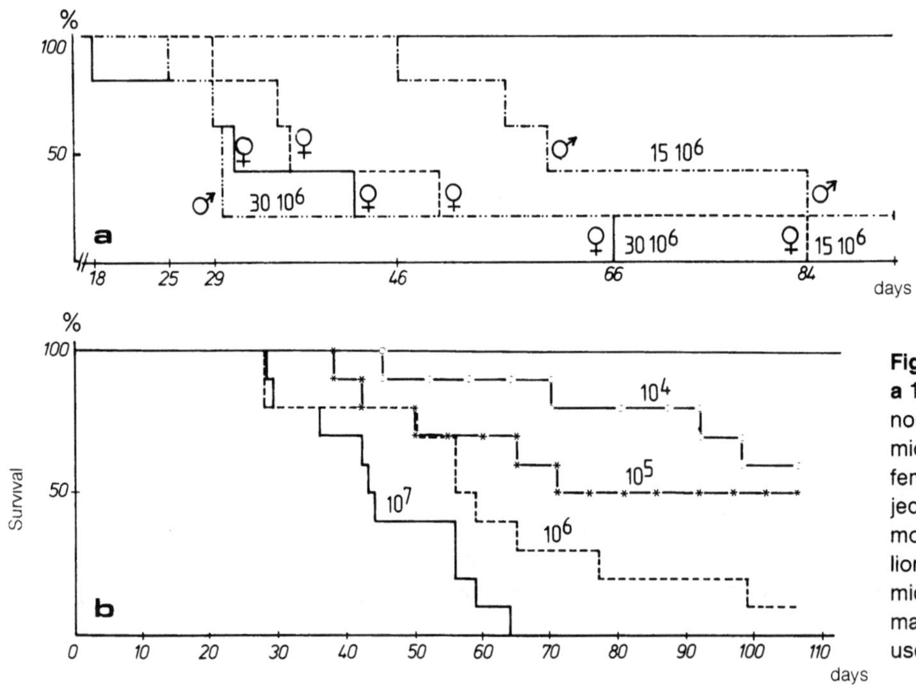

Fig. 1a, b. P-12 ICHIKAWA model. **a** 15 or 30 million cells injected i.p. to non-irradiated male and female mice. 100% death was only obtained with females. **b** different cell doses were injected i.v. into non-irradiated mice. 100% mortality was obtained only with 10 million cells, and with 1000 cells only 1/10 mice died on day 185 (not shown). Female nu/nu BALB/c BOM mice were used injected i.v. on day zero. On day 1, a single dose of only 20 μg of bound A-chain specific immunotoxin was injected, also i.v., alone or together with the potentiating agent HSA-SS-Monensin, at the dose of 250 μg of monensin. Controls received PBS or HSA-SS-Monensin alone. The association of immunotoxin (at a dose shown to be inefficient) and the potentiating agent HSA-SS-Monensin induced a significant prolongation of survival as compared to controls (50% survival observed on day 81 for the experimental group, versus on days 39-43 for the controls).

Other experiments were performed to study the best therapeutical protocol for the administration of this combined therapy. The same experimental model was used, with the cells injected i.v. 24 hours before the different compounds, which were also injected i.v. One of the problems found when injecting both products (immunotoxin and potentiator), was high mortality shortly after the injection, due to enhanced toxicity of the low doses of A-chain injected. In fact, the LD50 for immunotoxin, when given in conjunction with the potentiating agent, is 25 μg of A-chain. Therefore, different administration protocols were devised, in an attempt to avoid this non-specific early toxicity. Figure 3c summarizes the results obtained with the following treatment schedules: 10×10^6 cells were injected i.v. on day zero. Two experimental groups were tested: both received a low dose of immunotoxin (25 μg of bound A-chain) in association with the potentiating agent HSA-SS-Monensin, also i.v., but both molecules were injected separated by an interval of 16 hours: HSA-SS-Monensin was injected either 16 hours before or 16 hours after the immunotoxin injection. It can be seen that the best results are obtained when the immunotoxin is injected before the potentiating agent, suggesting that during that time all the target cells have been saturated with immunotoxin. Conversely, when the immunotoxin injection follows the potentiating agent, early deaths, due to the enhancing toxic effect of activated A-chain, are observed (3/10 mice on day 3 after treatment) and even if 50% survival is delayed, all animals die by day 118. Although confirmation has to be obtained, these results support the finding that immunotoxin can be very efficiently activated *in vivo*, and that the order in which the molecules are administered plays an important role. It can be expected that immunotoxin molecules are bound to the target cells during a sufficiently long time for the activating process to take place. Not shown in the figure is the control group, where the immunotoxin and the potentiating molecule were given together, the day after the cells, and 8/10 mice were dead 24 hours later. When the injections of both prod-

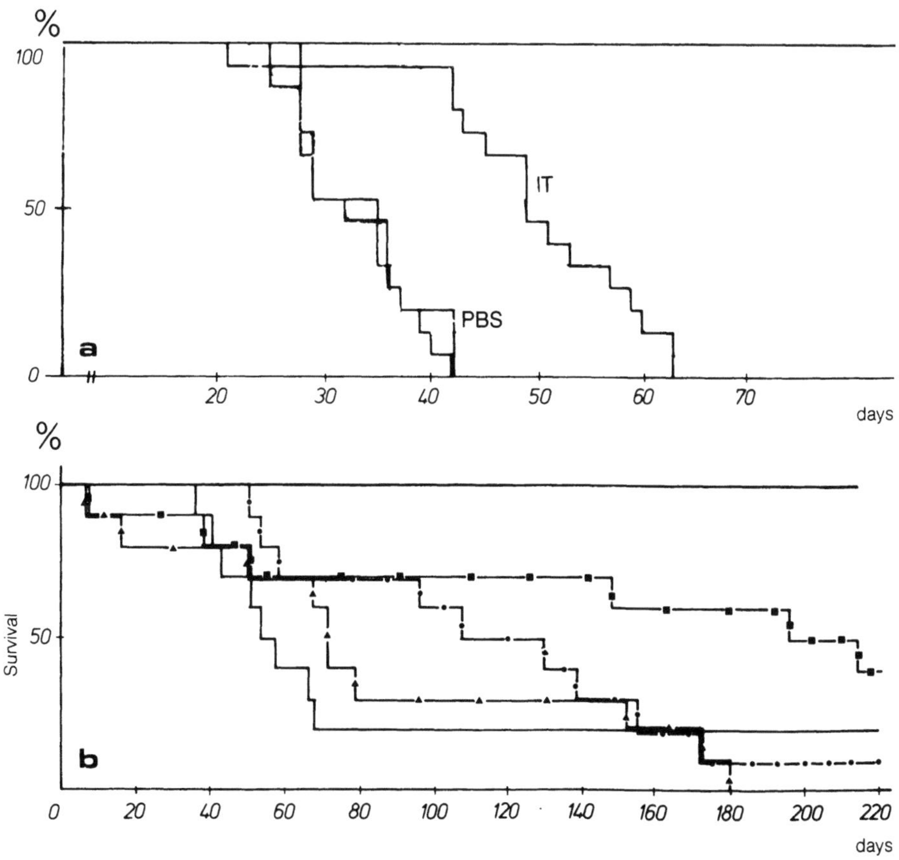

Fig. 2a, b. Treatment of the P12-ICHIKAWA human T-cell disease in nu/nu mice by specific immunotoxin. **a** 25 million cells were injected i.p. on day 0. Specific immunotoxin (T101) was injected, also i.p., on the 8 following days. **b** 10 million cells were injected i.v. on day 0. T101 - 400 μg bound A chain was i.p. administered on the 4 following days. (–) PBS; (– ▲ –) non-specific IT; (– ● –) T101 antibody; (– ■ –) T101 IT

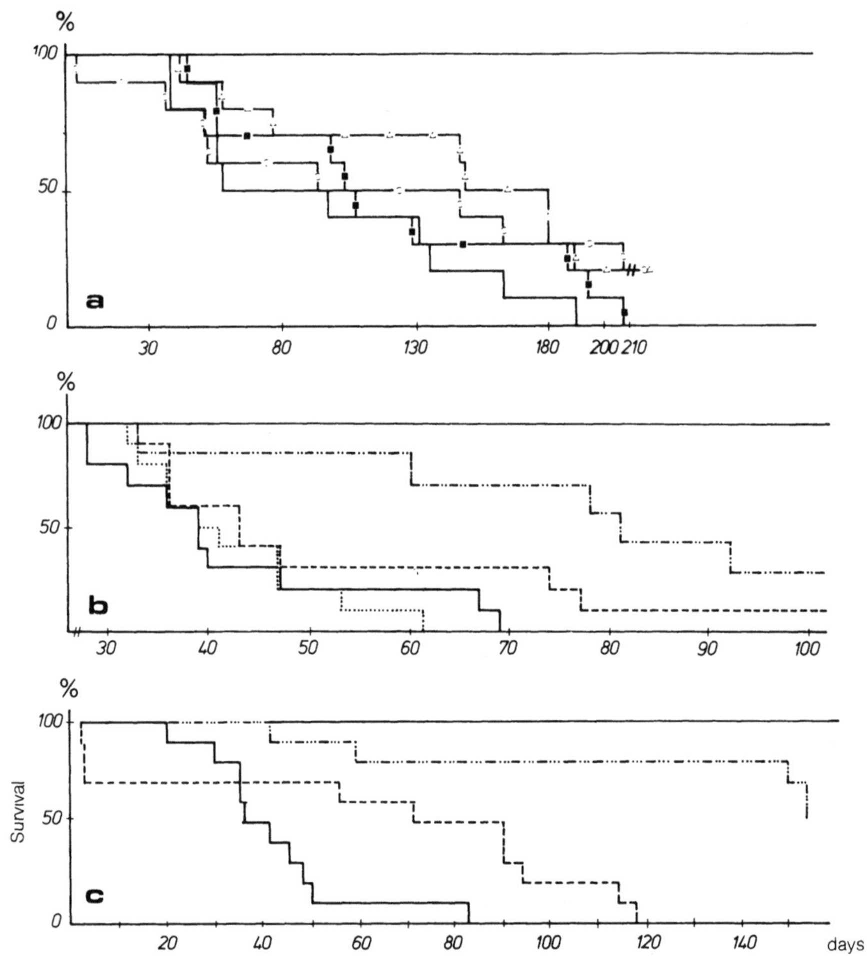

Figure 3a–c. Treatment of the P12-ICHIKAWA human T-cell disease in nu/nu mice by specific immunotoxin, associated with the potentiating agent HSA-SS-Monensin. **a** 10 million cells were injected i.v. on day zero. Treatment was also i.v., in a single dose of 200 μg of bound A-chain, given on day 1. No potentiating agent was added. (—) T101 antibody; (—■—) PBS; (—○—) non-specific IT; (—△—) T101 IT. **b** 10 million cells were injected i.v. on day zero. On day 1, a single dose of only 20 μg of bound A-chain specific immunotoxin was injected, also i.v., alone, or together with the potentiating agent HSA-SS-Monensin, at the dose of 250 μg of monensin. (—) PBS; (···) HSA-SS alone; (---) T101 IT alone; (-····-) T101 IT+HSA-SS-Monensin. **c** 10 million cells were injected i.v. on day zero. Two experimental groups were tested: both received a low dose of immunotoxin (25 μg of bound A-chain) in association with the potentiating agent HSA-SS-Monensin, also i.v., but both molecules were injected separately, at 16 hrs. interval (see text for details) (—) PBS; (---) HSA-SS-Monensin 16 hrs. before T101 IT; (-····-) HSA-SS-Monensin 16 hrs. after T101 IT

ucts were separated by only 8 hours, intermediate results were obtained. Our results clearly demonstrate that an immunotoxin can reach its target when injected into animals, even though optimal treatment protocols are to be defined for maximum efficacy.

Conclusion

Several human T-cell leukaemia cell lines, established in continuous culture by different laboratories, were tested for their potential to induce characterized disease in "nude" mice, for the purpose of studying the therapeutic potential of immunotoxins. Of all the lines tested, the only one giving reproducible results in our laboratory was the P12-ICHIKAWA line, bearing the differentiation antigen T65 or T1 (CD5) target for our T101-immunotoxin.

The advantage of this cell line is that even without irradiation of recipient mice, the disease can be obtained, even if relatively large numbers of cells are needed (10×10^6 for i.v. injection or 25×10^6 for i.p. administration).

This model has allowed us to demonstrate the therapeutic effect of a specific immunotoxin, the onset of the disease being delayed and even the number of dead mice being significantly reduced; it could also be demonstrated that low doses of immunotoxin, which are totally inefficient when injected alone, can be significantly enhanced when administered together with an activating molecule, the carboxylic ionophore monensin coupled to HSA as a carrier.

Although the extrapolation from these results to the clinical situation has to be carefully analysed, information is rapidly accumulating to support the use of these new hybrid molecules in the control and cure of malignant diseases.

Acknowledgements

Ichikawa cells were a generous gift of Dr. Shaw Watanabe, from the National Cancer Center Research Institute, Tokyo, Japan.

CEM cells were a generous gift of Dr. Ivor Royston, Veterans Administration Medical Center, La Jolla, San Diego, Ca., U.S.A.

The other cell lines mentioned were kindly provided to us by Dr. Stephan Carrel, Ludwig Institute, Lausanne, Switzerland; Dr. C. Rosenfeld, ICIG, Villejuif, France.

References

1. Watanabe S, Shimosato Y, Kameya T, Kuroki M, Kitahara T, Minato K and Shimoyama M: Leukemic distribution of a human acute lymphocytic leukemia cell line (Ichikawa strain) in nude mice conditioned with whole-body irradiation. Cancer Research 1978 (38): 3494-3498
2. Royston I, Majda JA, Baird SM, Meserve BL and Griffiths JC: Human T cell antigens defined by monoclonal antibody: The 65000 dalton antigen of T cells (T65) is also found on chronic lymphocytic leukemia cells bearing surface immunoglobulins. J Immunol 1980 (125): 725-731
3. Jansen FK, Blythman HE, Carriere D, Casellas P, Gros O, Gros P, Laurent JC, Paolucci F, Pau B, Poncelet P, Richer G, Vidal H and Voising GA: Immunotoxins: hybrid molecules combining high specificity and potent cytotoxicity. Immunological Rev 1982 (62): 185-216
4. Casellas P and Jansen FK: Enhancement of C-chain immunotoxins by pharmacological agents. In: Vogel CW (ed) Immunoconjugates: the current status of antibody conjugates for radioimaging and therapy of cancer. New York: Oxford University Press 1986

VII. Human Tumour Xenografts in Drug Screening

Proposal for the Application of Xenografts in Screening for New Anticancer Agents and in Selecting Tumor Types for Phase II Clinical Trials

B. WINOGRAD, M.W. LOBBEZOO, and H.M. PINEDO

EORTC New Drug Development Office, Free University Hospital, De Boelelaan 1117, 1081 HV Amsterdam, The Netherlands

Selection of Drugs for Clinical Evaluation: Preclinical Screening

The selection of drugs for clinical evaluation in cancer chemotherapy is usually based on the activity in several antitumor test systems. If a drug has a promising activity profile, can be produced in a reliable dosage form and does not show prohibitive toxicity in animals, it is a candidate for clinical evaluation. This is different from most other areas of drug development where a clinical candidate must perform well also in preclinical toxicology studies, thus having a large therapeutic index. In the past decades antitumor testing has mainly been performed in murine tumors *in vivo* and *in vitro*. A systematic screen to discover and characterize new lead compounds has been applied by the American National Cancer Institute (NCI) (1). Other institutes performing preclinical antitumor testing used the same or similar models and methods as the NCI. Although the NCI screening system was revised several times, the primary selection of new lead compounds depended heavily on the use of the murine leukemia models L1210 and P388, as the primary screen (2, 3). In 1975 several murine solid tumors and 3 human tumor xenografts (tumor lines) in nude mice were included in the tumor panel, i.e., secondary screen. These tumor models were expected to select drugs active in human solid tumors and predict clinical activity in the major human solid tumors, i.e., lung, breast and colorectal cancer (4, 5, 6).

Single Agent Activity: Clinical Phase I and Phase II Evaluation

A new drug is entered in phase I clinical trials to determine the maximum tolerated dose, the type of dose limiting toxicity and the human pharmacology at a given treatment schedule. A safe and optimal dose schedule should be proposed for phase II evaluation.

A common impression is that the therapeutic benefit during phase I trials is very small and therefore a therapeutic intent in these studies would be questionable. This has recently been demonstrated in a review with an overall response rate (partial and complete responses) of 4.2% (7). This low percentage might be due to the design of phase I trials with many patients being heavily pretreated, the majority receiving suboptimal doses and not all patients having measurable disease. However, response rates were demonstrated to be similarly low in a subset group of patients with measurable disease. Even for solid tumors commonly thought of as responsive and for drugs rated active after phase II evaluation, no better response rate was determined in phase I.

Single agent phase II studies are pivotal to assess the activity of a new drug against the different types of human cancer. Reviewing the literature of the past years regarding the response rate of new drugs against human solid tumors one will find an average of only 5% (data not summarized here). This percentage is in close agreement with the mean response rate of 4.5% which we calculated from recently published data for 47 new drugs in the following solid tumors: breast, colon, small cell lung, non-small cell lung and melanoma (8). Phase II studies are usually performed in a 'clinical panel', consisting of these 5 solid tumors plus adult acute leukemia and lymphoma (9). With only one exception, activity of the 47 new drugs in tumor types not represented in the 'clinical panel' always occurred in conjunction with activity against one or more of the panel tumors, thus indicating the reliability of this panel for broader antitumor activity. It should be noted, however, that a high percentage of new drugs rated as clinically active was active against lymphoma (17/23 or 74%) and against leukemia (7/20 or 35%) (8).

In view of the large number of patients treated with new drugs in phase II studies, the low response rate in solid tumors is very frustrating for the oncologist and new approaches to select candidate drugs for clinical studies in solid tumors seem necessary.

New Preclinical Screening Methods Needed

Drugs selected by the earlier screening method have been disappointing in their clinical activity. The sequential screening method with murine L1210 and P388 leukemias as major primary screen has selected several good drugs for the treatment of leukemias and lymphomas but has overall failed to identify drugs for the treatment of solid tumors. In general there was no correlation of activity between corresponding tumor histology in the murine systems of the NCI (secondary screen or tumor panel) and the clinic (2). The limited use of human tumor lines as xenografts did not contribute significantly in this respect.

It should be emphasized, however, that very few drugs were evaluated in the clinic when they failed to show activity in the above-mentioned sequential screening method. Nevertheless several of these drugs without activity in P388 and L1210 and some even without activity in the NCI tumor panel have shown significant clinical activity, i.e., bleomycin, deoxycoformycin, gallium nitrate, mitoguazone (MGBG).

In consequence, the NCI has chosen a new strategy for discovering lead compounds. A disease-oriented screen is being developed using a panel of human tumor cell lines *in vitro* as the primary screen (10).

Once new promising compounds have been discovered in early screening, it would be desirable to have a method which can select the solid tumor types in which a new drug should be evaluated clinically in phase II studies. The use of human tumor xenografts in nude mice (human tumor lines) might enable such a 'preclinical phase II' evaluation. This suggestion goes back to the years when human tumor xenografts began to be used as antitumor test model. If applied correctly this model best simulates clinical cancer. (In the following we will refer only to the 'subcutaneous assay' and not discuss results of the 'subrenal capsule assay'.)

Human Tumor Lines: Validation for Drug Screening

Prior to using a tumor model for drug screening it should be validated with respect to predictability for clinical activity. Such validation with chemotherapy can be done more thoroughly in human tumor lines than in murine tumors because original characteristics of a human tumor are well retained in the tumor line.

The first obvious question in validating a human tumor line is "How do drug effects in the nude mouse correlate with clinical results in the donating patient's tumor?" In a recent literature review we found in 79 patients 119 direct treatment comparisons between patients' treatment and laboratory data (Table 1). In 31/35 treatment comparisons the xenograft model predicted well for clinical response and in 79/84 for clinical progression. However, it should be noted that many of these positive correlations are based upon results obtained in clinically unresponsive tumor types (11).

In view of the variability in antitumor response to standard therapy between patients with the same tumor type, it is important to evaluate whether a panel of lines of the same tumor type can select the clinically relevant drugs for this tumor type and deselect drugs which only give low response rates. In analogy to clinical studies one might call this validation "preclinical phase II studies" (19). In a recent literature review (11) we found that prediction for clinical activity is better in clinically responsive tumors types, e.g. small cell lung cancer (20), ovarian cancer (21) and breast cancer (22, 23), than in clinically resistant tumor types, e.g. melanoma (17, 23, 24), colorectal carcinoma (12, 25–28) and non-small cell lung cancer (20, 23). Additional interesting results supporting the concept of preclinical phase II studies are presented in this monograph.

Preclinical Phase II Studies

Given the experience with chemotherapy studies in human tumor lines, showing good correlation between preclinical and clinical antitumor activity, we believe it is time to determine whether preclinical phase II studies in human tumor lines can improve the selection of drugs for disease-oriented clinical phase II studies. If the response rate in clinical phase II studies could be improved by this method, fewer patients would be exposed to aggressive experimental treatment without therapeutic benefit, and more active drugs could be evaluated in a better way.

For this approach several tumor types are to be chosen: tumor types clinically amenable to present chemotherapy, those unresponsive to present chemotherapy and intermediates. For each of these tumor types a representative panel of tumor lines will have to be selected. The number of lines per tu-

Table 1. Direct comparison of drug effects in the nude mouse with clinical results in the donating patient's tumor

Tumor type	No. pts	No. of treatment comparisons	Evaluation[a] criteria nude mouse	mouse/man[b]				Ref. no.
				R/R	P/P	R/P	P/R	
colorectal[c]	4	4	SGD > 2	1	3	-	-	12
lung SCLC[c]	6	8	SGD > 2	5	2	1	-	13
lung NSCLC[c]	9	12	SGD > 2	-	12	-	-	13
colorectal	1	1	MTW T/C > 75%	1	-	-	-	14
breast	1	1	MTW T/C > 75%	-	1	-	-	14
gastric	2	3	MTW T/C > 75%	-	3	-	-	14
colorectal	14	22	PR + CR	2	19	-	1	15
gastric	6	9	PR + CR	4	5	-	-	15
lung NSCLC	4	6	PR + CR	6	-	-	-	15
lung SCLC	2	4	PR + CR	-	4	-	-	16
melanoma	4	4	PR + CR	-	3	1	-	15
sarcoma STS	2	2	PR + CR	-	2	-	-	15
penile	1	1	PR + CR	1	-	-	-	15
Wilms	1	2	PR + CR	2	-	-	-	15
thyroid	1	1	PR + CR	1	-	-	-	15
melanoma	17	37	GD > 2	8	25	3	1	17
head + neck	2	2	GD > 2	-	-	-	2	18
Total	79	119		31	79	5	4	

[a] SGD = specific growth delay; MTW = mean tumor weight treated/control; CR = complete response; PR = partial response; GD = growth delay.
[b] Comparison of drug effects nude mouse/patient.
R = clinical response (CR + PR) or meets evaluation criteria in the nude mouse.
P = clinical progression or does not meet evaluation criteria in the nude mouse.
[c] Tumor lines in artificially immune-deprived mice.

mor type will depend on the number of relevant characteristics needed to make the panel representative for the given tumor type. For each selected tumor line a fingerprint of histological, immunohistochemical, immunological and biochemical characterizations should be established in order to demonstrate its close similarity to the original tumor tissue in the patient and to confirm the retention of the characteristics through serial passages. Comparison of sensitivity to clinically useful drugs and resistance to clinically inactive drugs will help to select a panel of tumor lines representing the variability in antitumor response to standard therapy between patients with the same tumor type.

It is concluded that the value of panels of human tumor lines as subcutaneous xenografts in nude mice for the prediction of drug efficacy in phase II clinical trials should be investigated in a retrospective and prospective manner.

References

1. Venditti JM: Preclinical drug development: rationale and methods. Sem in Oncol 1981 (8): 349-361
2. Venditti JM, Wesley RA and Plowman J: Current NCI preclinical antitumor screening in vivo: results of tumor panel screening 1976-1982 and future directions. In: Garattini S, Goldin A, Hawkins S (eds) Advances in Pharmacology and Chemotherapy. Academic Press, Orlando 1984 (20): 1-20
3. Goldin A, Venditti JM, MacDonald JS, Muggia FM, Henney JE and DeVita Jr VT: Current results of the screening program at the Division of Cancer Treatment, National Cancer Institute. Eur J Cancer 1981 (17): 129-142
4. Corbett TH, Griswold jr DP, Roberts BJ, Peckham JC and Schabel Jr FM: Evaluation of single agents and combinations of chemotherapeutic agents in mouse colon carcinomas. Cancer 1977 (40): 2660-2680
5. Martin DS, Fugmann RA, Stolfi RL and Hayworth PE: Solid tumor animal model therapeutically predictive for human breast cancer. Cancer Chemother Rep Part 2 1975 (5): 89-109
6. Ovejera AA and Houchens DP: Human tumor xenografts in athymic nude mice as a preclinical screen for anticancer agents. Sem in Oncol 1981 (8): 386-393
7. Estey E, Hoth D, Simon R, Marsoni S, Leyland-Jones B and Wittes R: Therapeutic response in phase I trials of antineoplastic agents. Cancer Treat Rep 1986 (70): 1105-1115
8. Marsoni S, Hoth D, Simon R, Leyland-Jones B, De Rosa M and Wittes RE: Clinical drug development: an analysis of phase II trials, 1970-1985. Cancer Treat Rep 1987 (71): 71-80
9. DeVita VT, Oliverio VT, Muggia FM et al: The drug development and clinical trials programs of the Division of Cancer Treatment, National Cancer Institute. Cancer Clin Trials 1979 (2): 195-216
10. DCT board approves New Screening Programme, natural product concepts. The Cancer Letter 1985 (11): 4-5

11. Winograd B, Boven E, Lobbezoo MW and Pinedo HM: Human tumour xenografts in the nude mouse and their value as test models in anticancer drug development (review). In Vivo 1987 (1): 1-14
12. Nowak K, Peckham MJ and Steel GG: Variation in response of xenografts of colo-rectal carcinoma to chemotherapy. Br J Cancer 1978 (37): 576-584
13. Shorthouse AJ, Smyth JF, Steel GG, Ellison M, Mills J and Peckham MJ: The human tumor xenograft - a valid model in experimental chemotherapy? Br J Surg 1980 (67): 715-722
14. Fujita M, Hayata S and Taguchi T: Relationship of chemotherapy on human cancer xenografts in nude mice to clinical response in donor patients. J Surg Oncol 1981 (15): 211-219
15. Fiebig HH, Schuchhardt C, Henss H, Fiedler L and Löhr GW: Comparison of tumor response in nude mice and in the patients. Behring Inst Mitt 1984 (74): 343-352
16. Fiebig HH, Neumann HA, Henss H, Koch H, Kaiser D and Arnold H: Development of three human small cell lung cancer models in nude mice. Res Results Cancer Res 1985 (97): 77-86
17. Osieka R: Studies on drug resistance in a human melanoma xenograft system. Cancer Treat Rev 1984 (11, Suppl A): 85-97
18. Braakhuis BJM, Schoevers EJ, Heinerman ECM, Sneeuwloper G and Snow GB: Chemotherapy of human head and neck cancer xenografts with three clinically active drugs: cis-platinum, bleomycin and methotrexate. Br J Cancer 1983 (48): 711-716
19. Bellet RE, Danna V, Mastrangelo MKJ and Berd D: Evaluation of a "nude" mouse-human tumor panel as a predictive secondary screen for cancer chemotherapeutic agents. JNCI 1979 (63): 1185-1188
20. Shorthouse AJ, Jones JM, Steel GG and Peckham MJ: Experimental combination and single-agent chemotherapy in human lung-tumour xenografts. Br J Cancer 1982 (46): 35-44
21. Boven E: Conventional agents in human ovarian cancer xenografts. In: Winograd B, Peckham MJ, Pinedo HM (eds) Human Tumour Xenografts in Anticancer Drug Development. ESO Monograph Series. Springer Verlag, Heidelberg 1987: 33-35
22. Bailey MJ, Gazet JC, Smith IE and Steel GG: Chemotherapy of human breast-carcinoma xenografts. Br J Cancer 1980 (42): 530-536
23. Giuliani FC, Zirvi KA and Kaplan NO: Therapeutic response of human tumor xenografts in athymic mice to doxorubicin. Cancer Res 1981 (41): 325-335
24. Fodstad Ø, Aass N and Pihl A: Response to chemotherapy of human malignant melanoma xenografts in athymic, nude mice. Int J Cancer 1980 (25): 453-458
25. Fiebig HH, Widmer KH, Fiedler L, Wittekind O and Löhr GW: Development and characterization of 51 human tumor models for large bowel, stomach and esophageal cancer. Dig Surg 1984 (1): 225-235
26. Giuliani FC, Zirvi KA, Kaplan NO and Goldin A: Chemotherapy of human colorectal tumor xenografts in athymic mice with clinically active drugs: 5-fluorouracil and 1-3-bis-(-2-chloroethyl)-1-nitrosourea (BCNU). Comparison with doxorubicin derivatives: 4′-deoxydoxorubicin and 4′-0-methyldoxorubicin. Int J Cancer 1981 (27): 5-13
27. Osieka R, Houchens DP, Goldin A and Johnson RK: Chemotherapy of human colon cancer xenografts in athymic nude mice. Cancer 1977 (40): 2640-2650
28. Houghton PJ and Houghton JA: Evaluation of single agent therapy in human colorectal tumour xenografts. Br J Cancer 1978 (37): 833-840

Human Tumour Xenograft Models for Use with an In Vitro-Based, Disease-Oriented Antitumour Drug Screening Program

R. H. SHOEMAKER[1]*, T. L. MCLEMORE[1], B. J. ABBOTT[1], D. L. FINE[2], E. GORELIK[3], J. G. MAYO[1], Ø. FODSTAD[4], and M. R. BOYD[1]

[1] Developmental Therapeutics Program, Division of Cancer Treatment, National Cancer Institute, Bethesda, MD 20892, U.S.A.
[2] Program Resources, Inc., National Cancer Institute - Frederick Cancer Research Facility, Frederick, MD 21701, U.S.A.
[3] Pittsburgh Cancer Institute, 230 Lothrop Street, Pittsburgh, PA 15213, U.S.A.
[4] Institute for Cancer Research, Montebello, 0310 Oslo 3, Norway

Abstract

Both short-term and long-term xenograft models may be useful in conjunction with an *in vitro* based disease-oriented drug screening program. Short-term models may be most valuable in making initial assessments of the potential *in vitro* drug screening leads for *in vivo* use. We have previously shown that a substantial number of such leads may be subject to metabolic inactivation (1) and that this may be associated with a lack of therapeutic activity and a relative lack of toxicity *in vivo*. Certainly, rapid excretion or other pharmacologic factors may also render compounds inactive *in vivo*. Short-term assays may be very useful for identifying such compounds and thus setting priorities for further testing of *in vitro* drug leads in more rigorous longer-term models. Rational application of these longer-term models with particular attention to modeling of *in situ* vascular barriers, tumor microenvironment, and the natural history of the target diseases may facilitate identification and development of new drugs with significant clinical activity against the common adult solid tumors.

Introduction

Human tumor xenografts have been widely used for experimental therapeutic studies including new drug testing. Since 1975 the National Cancer Institute (NCI) of the United States has utilized a group of 3 xenografts as a component of a "Tumor Panel" for evaluation of compounds active in a primary *in vivo* drug screen based on the murine P388 leukemia model (2, 3, 4). While this Tumor Panel experiment has identified some compounds with substantial activity against one or more of the xenograft models, the reliance of this screening strategy on P388 as a primary screen has been thought to severely limit its potential for identifying new drugs with selective activity towards human solid tumors. Indeed, the clinical activity of compounds identified in the P388 based screen has been limited mainly to leukemias and lymphomas. We have recently begun development and implementation of a new "disease-oriented" drug screening strategy based on the use of relatively large panels of human tumor cell lines representing various tumor (disease) types (5, 6). In this new strategy each compound to be screened for antitumor activity will be tested *in vitro* in each member of the cell line panels. Information on the relative sensitivity of tumor cell lines (e.g. IC_{50}; the concentration of test compound resulting in inhibition of growth to 50% of the control value) will be the basis for identification of compounds exerting differential cytotoxicity (significant differences in IC_{50}s) and, within this category, identification of compounds showing selective cytotoxicity to particular cell types. These *in vitro* drug screening leads will require follow-up testing *in vivo* and this may most appropriately be performed using the cell lines which are most sensitive *in vitro*. Since a large number of cell lines with diverse properties will be routinely used for screening, this strategy poses some unique technical and logistic problems. In this report we shall briefly describe some of the *in vivo* models currently being investigated for possible use in this program and highlight some of the relative advantages and disadvantages of these models.

Human Tumor Xenograft Models

Human tumor xenograft models can be generally categorized as long-term or short-term assays. The "classical" long-term xenograft model, which evolved quite naturally out of the initial surveys of tumorigenicity of various human neoplasms in ei-

* To whom reprint requests should be addressed: National Cancer Institute - Frederick Cancer Research Facility, Building 428, Room 63, Frederick, MD 21701, U.S.A.

ther artificially immune suppressed or genetically immunodeficient (nude) mice and the conventional practices applied with murine solid tumors, involved the serial measurement of subcutaneous tumor masses under various treatment conditions (7–8). This model is amenable to precise measurement of drug effects (10), has analogy to clinically utilized parameters of response (11), and allows systemic drug administration via either the intravenous or intraperitoneal routes. This latter route is of practical advantage for newly identified compounds which may not be readily formulated for intravenous administration. A very wide variety of histologic types of human tumors have been shown to form progressively growing subcutaneous tumors in nude mice (12). For the most part, however, tumors transplanted to the subcutaneous site show a relatively benign pattern of growth, i.e., tumors expand locally and become associated with a dense connective tissue capsule. Typically, these tumors do not metastasize in the nude mouse (13, 14) and if they are lethal it is due to ulceration, bleeding, infection, nutritional depletion, or mechanical interference with the functioning of the animal (e.g. access to water, food, etc.). In spite of this paradoxically benign pattern of growth, the subcutaneous tumor growth inhibition model seems to reflect some relevant features of the original patient's tumor *in situ.* From the standpoint of new drug screening the evidence for expression of drug sensitivity/resistance patterns corresponding to the clinical response patterns provides the best indication for the validity of this model (11, 15, 16, 17).

Several important practical problems arise in the large-scale use of subcutaneous assays for drug testing. Some human tumors grow very slowly and may require more than 6 months for completion of an assay. Growth inhibition assays are rather labor intensive, optimally involving multiple individual tumor measurements and requiring maintenance of specific-pathogen-free conditions. Taken together, these factors result in a rather high cost per assay and long turn-around time for assay results. Furthermore, some types of human tumor cells, e.g. lymphomas, tend not to form subcutaneous tumors in nude mice (18) and therefore cannot be readily studied with this model.

Short-Term Xenograft Models

The practical problems associated with subcutaneous tumor growth inhibition assays have led to attempts to develop short-term assays which can be performed rapidly, relatively inexpensively, and with all or a majority of tumor cell types of interest.

The subrenal capsule assay (SRCA) developed by Bogden et al. (19, 20) meets the first two of these objectives reasonably well and can be applied to a wide range of tumor types. As mentioned above, NCI has made extensive use of the SRCA in secondary evaluation of P388 active compounds. This testing, however, involved only 3 different xenografts (1 breast, 1 colon, and 1 lung carcinoma). We have recently begun development of protocols for SRCA using the large number of cell lines acquired and characterized for use in the *in vitro* based, disease-oriented screening program (Table 1). All of

Table 1. Human tumor cell lines acquired for use in disease-oriented antitumor drug screening[a]

NON-SMALL CELL LUNG CANCER		
Adenocarcinoma	*Adeno-squamous Carcinoma*	*Squamous Cell Carcinoma*
NCI-H23	NCI-H125	NCI-H520
NCI-H522		
A549/Asc-1		
EKVX		
Bronchiolo-alveolar Carcinoma	*Large Cell Carcinoma*	
NCI-H322M	NCI-H460	
NCI-H358M		
SMALL CELL LUNG CANCER (SCLC)		
"Classic" SCLC	*"Variant" SCLC*	*"Adherent" SCLC*
NCI-H69	NCI-H82	DMS 114
NCI-H146	NCI-H524	
COLON CANCER	MELANOMA	RENAL CANCER
HT-29	LOX	UO-31
LoVo	Malme-3M	SN12 C
SW620	RPMI-7951	SN12 K1
DLD-1	SK-MEL-2	A498
	SK-MEL-5	A704
OVARIAN CANCER	CNS CANCER	PROSTATE CANCER
OVCAR 3	SNB19	PC-3
OVCAR 4	SNB44	PC-3M
OVCAR 5	SNB175	DU-145
OVCAR 8	U251	LNCAP
A2780	TE671	1013 L
LEUKEMIA	MISCELLANEOUS CELL LINES	
CCRF-CEM	P388 (Murine Leukemia)	
CCRF-SB	P388/Adriamycin resistant	
K-562	MCF-7 (Breast Carcinoma)	
Molt-4	MCF-7/Adriamycin resistant	
HL-60	CCD-19Lu (Lung Fibroblast)	

[a] This listing reflects cell lines initially acquired and characterized for use in disease-oriented drug screening. Thus, this table does not represent a fixed screening panel but rather, an initial panel which will evolve as additional or improved cell lines become available.

the lines which we have worked with to date have been amenable to use in the SRCA. However, protocols for individual tumors have required tailoring to suit their individual growth characteristics (Table 2). The SRCA and, in particular, an adaptation of the assay which employs immunocompetent mice in an abbreviated (six-day) assay, has been the subject of considerable criticism (21, 22, 23, 24). It has been our experience that when sterile, viable tumor fragments are employed and assays carried out as described in Table 2 using immunodeficient (nude) mice maintained under pathogen-free conditions, the SRCA can give reproducible measurements of antitumor effects with these tumor cells. However, the requirement to initially grow the tumor cells in the subcutaneous site to provide a source of solid tumor fragments for implantation under the renal capsule still contributes to the complexity, cost, and turn-around time for the SRCA. Thus, while an individual assay may be completed in as little as 6 days, when the lead time necessary to establish the subcutaneous tumors is included, it can take several months to get an individual tumor "on-line" for use in drug testing. This is of considerable significance to our current screening strategy where it would be prohibitively expensive to maintain more than 50 tumors in continuous xenograft passage. The recent demonstration by Fingert et al. (25) that cells harvested from cell culture flasks, or even freshly dissociated tumor cells from surgical specimens, can be placed in a fibrin clot and used directly in the SRCA merits further evaluation as a potential solution to this problem.

Table 2. Treatment schedule and assay evaluation day for human tumor xenograft subrenal capsule protocols under development

TUMOR CELL LINE	HISTOLOGIC TYPE	TREATMENT SCHEDULE	EVALUATION DAY
LOX	Melanoma	Q4D × 2 (days 1, 3)	day 6
HT-29	Colon adenocarcinoma	Q4D × 4 (days 1, 5, 9, 13)	day 15
NCI-H23	Lung adenocarcinoma	Q4D × 4 (days 5, 9, 13, 17)	day 18
A549	Lung adenocarcinoma	Q4D × 3 (days 1, 5, 9)	day 12
NCI-H125	Lung adenosquamous carcinoma	Q4D × 3 (days 5, 9, 13)	day 16
NCI-H322M	Bronchioloalveolar carcinoma	Q4D × 4 (days 4, 8, 12, 16)	day 18
NCI-H358M	Bronchioloalveolar carcinoma	Q4D × 3 (days 5, 9, 13)	day 16
NCI-H460	Lung large cell carcinoma	Q4D × 4 (days 1, 5, 9, 13)	day 15
NCI-H69	Lung small cell carcinoma	Q4D × 3 (days 4, 8, 12)	day 14

We have recently begun to explore the use of microencapsulation technology as an alternative approach to development of a short-term *in vivo* assay (26, 27). Using a process developed by Damon Biotech (Boston, MA), tumor cells harvested from actively growing cultures or cryopreserved cells are trapped in small (approximately 1 mm in diameter) spherical capsules with semipermeable membranes. These microcapsules are then implanted into the peritoneal cavity of nude mice. Drug treatments may then be applied by the intraperitoneal or intravenous routes, capsules recovered, and treatment effects measured on the basis of tumor cell numbers collected from disrupted capsules. While our experience with this approach is still rather limited, we have observed that all tumor cells tested to date (more than 10 different lines representing 6 histologic types), grow well in this "Micro-Encapsulated Tumor Assay" (META) and that relative drug sensitivity in this assay corresponds closely to that obtained with long-term *in vivo* assays. As in the SRCA the duration of the assay must be customized for individual cell lines but for all those examined to date the META can be completed in less than 3 weeks. Thus, it appears that the META may be particularly suitable for use as an initial short-term *in vivo* assay for evaluation of *in vitro* drug screening leads.

Long-Term Xenograft Models

While short-term assays such as the SRCA or META may provide valuable information regarding the compatibility of a particular compound with *in vivo* administration e.g., establish bioavailability under conditions of pharmacologic processing, longer-term assays provide an opportunity for more rigorous evaluation. As mentioned above, the classical subcutaneous tumor growth inhibition assay may be used in this context and allows precise measurement of therapeutic effects. Subcutaneous implantation, however, does not correspond to the anatomic location of the common solid tumors in the clinic and thus fails to model factors such as blood supply and microenvironment. Furthermore, there is no opportunity in this model to assess the effect of compounds on the process of tumor metastasis or on treatment of established metastases. In order for systemic chemotherapy to effect cure it is essential that metastases be effectively treated (28, 29). To address these issues we have begun development of orthotopic transplantation models (30, 31) and metastatic models (32).

Fidler (33) has argued the importance of transplantation of relevant tumor cell populations to relevant anatomic sites for experimental therapeutic studies and, in particular, for studies of metastasis. The use of orthotopic transplantation, i.e. implantation of human tumors in nude mice at sites corresponding to the site of tumor origin, was first reported for chemotherapy of brain tumors (34). Human glioblastoma (U-251) or medulloblastoma (TE-671) cells implanted intracerebrally in nude mice grew progressively, killing the mice and producing reproducible death patterns consistent with the use of survival as endpoint for experimental therapeutic trials. These brain tumor models have proven to be relatively refractory to chemotherapy, particularly the medulloblastoma model, but do show some therapeutic response to agents such as BCNU which are clinically used in brain cancer treatment. The intracerebral models allow preclinical testing of compounds in an *in vivo* environment where vascular impediments to drug delivery ("blood-brain barrier") and microenvironmental factors can be represented which may be of relevance to the potential clinical activity of the compound in patients with brain tumors.

Because of the increasing public health problem posed by lung cancer (35) and the lack of effective chemotherapy of this disease (36), we have focused our initial *in vitro* and *in vivo* model development efforts in this area. *In vivo* lung cancer models are needed which represent various stages of primary lung cancer, as well as models of metastatic cancer in the lung. As an approach to the former problem we have explored the orthotopic transplantation of lung cancer cell lines. Recently we have found that established human lung cancer cell lines grow readily within the lung parenchyma and/or pleural space following intrathoracic implantation (via percutaneous inoculation directly into the lung) or following intrabronchial (i.b.) instillation of tumor cell suspensions (via tracheal cut-down and insertion of a catheter into the right mainstem bronchus) (30, 31). We have further observed that cell suspensions obtained by enzymatic disaggregation of fresh lung tumor surgical specimens also grow readily following i.b. implantation. Tumors are formed within the right lung which grow progressively, involving the mediastinal lymph nodes, displacing the heart and left lung, and ultimately result in death of the animals. A variety of lung cancer cell lines have been shown to produce reproducible death patterns (median survival times of 11–120 days) following i.b. inoculation of 1×10^6–1×10^5 cells. We have recently begun experimental chemotherapy trials with the i.b. model using increase in lifespan as assay endpoint.

In the course of experiments aimed at development of metastatic human tumor models we have developed several tumor cell lines which show particularly aggressive growth in nude mice, several of these, most notably the LOX amelanotic melanoma (37, 38), produce lethal disease following intraperitoneal (i.p.) implantation. Intraperitoneal survival models have also been developed by other groups beginning with the ovarian carcinomatosis model of Ozols et al. (39). These survival models have the advantage of using an unambiguous endpoint (death), are readily evaluated in terms of fraction of animals cured or "log cell kill", and require relatively little manipulation of the animals. In the case of the LOX i.p. model we have observed that tumor cells spread rapidly (in as little as 1 day) from the peritoneal cavity to the lung and form metastatic lesions within lung parenchyma. Thus, by staging animals for several days prior to initiation of therapy, a model is obtained in which curative treatment requires eradication of the pulmonary metastases.

Several investigators have observed metastasis of human tumors in nude mice (40, 41, 42, 43) and in some cases it has been possible, through selection, to derive human tumor cell populations with enhanced metastatic potential (43). We have been primarily interested in the development of practical metastatic models based on the human tumor cell lines selected for use in our *in vitro* drug screening panel (32). For such a model to be practically useful it is critical that metastasis be observed reproducibly. Thus far we have been successful in establishing one practical metastatic model for use in "routine" drug evaluation. In this model, a selected metastatic line of the LOX melanoma is implanted subcutaneously, tumors are staged for 12 or more days to allow pulmonary metastases to become established, and then treatment is administered (generally by the i.v. route to simulate systemic chemotherapy in the clinic). The animals are killed one or more days later and lung fragments are transplanted subcutaneously to "bioassay" animals. The effect of chemotherapy on pulmonary metastases is assessed on the basis of inhibition of tumor outgrowth in bioassay animals receiving lung fragments from treated compared to control animals. With this LOX line, 100 percent of control animals develop progressively growing subcutaneous tumors (and a new generation of pulmonary metastases). In the case of melanoma this bioassay approach allows reasonable modelling of the clinical anatomic location of the primary tumor and one of its principal patterns of metastasis. Drugs showing selective cytotoxicity to melanoma metastatic model such as this, would be considered promising candidates for development to clinical trial.

Acknowledgements

The authors thank Dr. Daniel Griswold and Mr. Donald Dykes of the Southern Research Institute for assistance in development of the metastatic melanoma model and Mrs. Kathy Gill for preparation of this manuscript. We would also like to thank the following individuals for providing samples of the cell lines described in this report: Dr. Isiah Fidler (HT-29, A549, SN12C and SN12 K1); Dr. Kenneth Cowan (MCF-7 and MCF-7/Adriamycin Resistant); Dr. Paul Kornblith (SNB19, SNB44 and SNB75); Dr. M. Edward Kaighn (PC-3); Dr. Marsden Linehan (UO-31); Drs. John Minna, Adi Gazdar and Herbert Oie (lung cancer cell lines designated NCI-H...); Drs. Robert Ozols and Thomas Hamilton (OVCAR lines); and Dr. Olive Pettengill (DMS 114). Samples of other cell lines were obtained from the American Type Culture Collection, Rockville, MD, USA.

References

1. Appel PL, Alley MC, Lieber MM, Shoemaker RH and Powis G: Metabolic stability of experimental chemotherapeutic agents in hepatocyte: tumor cell co-cultures. Cancer Chemother Pharmacol 1986 (17): 47-52
2. Venditti JM: Preclinical drug development: Rationale and methods. Semin Oncol 1981 (8): 349-361
3. Staquet MJ, Byar DP, Green SB and Rozencweig M: Clinical predictivity of transplantable tumor systems in the selection of new drugs for solid tumors: Rationale for a three-stage strategy. Cancer Treat Rep 1983 (67): 753-765
4. Venditti JM, Wesley RA and Plowmann J: Current NCI preclinical screening in vivo results of tumor panel screening 1976-1982 and future directions. In: Garattini S, Goldin A and Hawkings S (eds) Advances in Pharmacology and Chemotherapy. Orlando, FL: Academic Press 1984: 1-20
5. Boyd MR, Shoemaker RH, Cragg GM and Suffness M: Future perspectives on the anticancer drug discovery program of the National Cancer Institute: Potential avenues for the investigation of marine biologicals. In: Jefford CW, Reinhart KL, Shield LS (eds) Pharmaceuticals and the Sea. Technomic Publ AG, Basel, in press
6. Boyd MR, Shoemaker RH, Alley MC, Scudiero DA, Monks A, Fine DL, Mayo JG and Chabner BA: Drug development. In: Roth JA, Ruckdeschel JC and Weisenburger TH (eds) Thoracic Oncology. W. B. Saunders Co. In press
7. Geran RI, Greenberg NH, MacDonald MM, Schumacher AM and Abbott BJ: Protocols for screening chemical agents and natural products against animal tumors and other biological systems. Cancer Chemother Rep 1972 (3): 1-103
8. In vivo cancer models. 1984 NIH Publication No. 84-2635. US Department of Health and Human Services, Public Health Service, National Institutes of Health, Bethesda, MD
9. Houghton JA, Cook RL, Lutz PJ and Houghton PJ: Childhood rhabdomyosarcoma xenografts: Responses to DNA-interacting agents and agents used in current clinical therapy. Eur J Cancer Clin Oncol 1984 (20): 955-960
10. Steel GG, Courtenay VD and Peckham MJ: The response to chemotherapy of a variety of human tumor xenografts. Br J Cancer 1983 (47): 1-13
11. Fiebig HH, Schuchhardt C, Henss H, Fiedler L and Löhr GW: Comparison of tumor response in nude mice and in the patients. Behring Inst Mitt 1984 (74): 343-352
12. Fogh J, Fogh JM and Orfeo T: One hundred and twenty-seven cultured human tumor cell lines producing tumors in nude mice. J Natl Cancer Inst 1977 (59): 221-226
13. Sordat B, Fritsche R, Mach P, Carrel S, Ozello L and Cerotini JC: Morphological and functional evaluation of human solid tumors serially transplanted in nude mice. In: Rygaard J and Povlsen CO (eds) Proceedings of the first international workshop on nude mice. Stuttgart: Fischer 1974: 269-277
14. Sharkey FE and Fogh J: Metastasis of human tumors in athymic nude mice. Int J Cancer 1979 (24): 733-738
15. Bellet RE, Danna V, Mastrangelo MJ and Berd D: Evaluation of a nude mouse-human tumor panel as a predictive secondary screen for cancer chemotherapeutic agents. J Natl Cancer Inst 1979 (63): 1185-1188
16. Bogden AE, Griffin W, Reich SD, Costanza ME and Cobb WR: Predictive testing with the subrenal capsule assay. Cancer Treat Rev 1984 (11): 113-124
17. Giovanella BC, Stehlin JS Jr, Shepard RC and Williams LJ: Correlation between response to chemotherapy of human tumors in patients and in nude mice. Cancer 1983 (52): 1146-1152
18. Epstein AL, Herman MM, Kim H, Dorfman FR and Kaplan HS: Biology of the human malignant lymphomas. Cancer 1976 (37): 2158-2176
19. Bogden AE, Kelton DE, Cobb WR and Esber HJ: A rapid screening method for testing chemotherapeutic agents against human tumor xenografts. In: Houchens D and Ovejera T (eds) Proceedings of the symposium on the use of athymic (nude) mice in cancer research. New York: Gustav Fischer 1978
20. Bogden AE, Haskell PM, LePage DJ, Kelton DE, Cobb WR and Esber HJ: Growth of human tumor xenografts implanted under the renal capsule of normal immunocompetent mice. Exp Cell Biol 1979 (47): 281-293
21. Edelstein MB, Smink T, Ruiter DJ, Visser W and Van Putten LM: Improvements and limitations of the subrenal capsule assay for determining tumor sensitivity to cytostatic drugs. Eur J Cancer Clin Oncol 1984 (12): 1549-1556
22. Edelstein MD, Fiebig HH, Smink T, Van Putten LM and Schuchhardt C: Comparison between macroscopic and microscopic evaluation of tumor responsiveness using the subrenal capsule assay. Eur J Cancer Clin Oncol 1983 19 (7): 995-1009
23. Aamdal S, Fodstad Ø, Kaalhus O and Pihl A: Chemosensitivity profiles of human cancers assessed by the 6-day SRC assay on serially xenografted tumors. Int J Cancer 1986 (37): 579-587

24. Bennett JA, Pilon VA and MacDowell RT: Evaluation of growth and histology of human tumor xenografts implanted under the renal capsule of immunocompetent and immunodeficient mice. Cancer Res 1985 (45): 4963-4969
25. Fingert HJ, Mizrahi N, Bamberg MP and Cohen AM: Histologic studies of human tumors in the subrenal capsule assay (SRCA): Improved growth by enzymatic disaggregation. Proc Amer Assoc Cancer Res 1986 (27): 384
26. Gorelik E, Alley M and Shoemaker R: A new in vivo short-term assay for evaluation of antitumor chemotherapeutic drugs. Proc Amer Assoc Cancer Res 1986 (27): 389
27. Gorelik E, Shoemaker R, Jarvis A, Ovejera A, Alley M, Duff R, Mayo J and Boyd M: A new short-term assay for in vivo evaluation of the effect of anticancer drugs on human tumor cells. (Manuscript in preparation)
28. Fidler IJ: General considerations for studies of experimental cancer metastasis. Meth Cancer Res 1978 (15): 399-439
29. Fidler IJ: Recent concepts of cancer metastasis and their implications for therapy. Cancer Treat Rep 1984 (68): 193-198
30. McLemore TL, Blacker PC, Gregg M, Jessee SE, Alley MC, Abbott BJ, Shoemaker RH, Litterst CC, Hubbard WC, Brennan RN, Fine DL, Eggleston JC, Mayo JG and Boyd MR: Intrabronchial implantation: A method for the orthotopic propagation of human lung tumors in athymic nude mice. Chest 1987 (91): 55-85
31. McLemore TL, Blacker PC, Gregg M, Alley MC, Abbott BJ, Shoemaker RH, Liu MC, Litterst CC, Hubbard WC, Brennan RN, Fine DL, Bohlman ME, Eggleston JC, Mayo JG and Boyd MR: A novel intrapulmonary model for the orthotopic propagation of human lung cancers in athymic nude mice. Cancer Res 1987 (47): 5132-5140
32. Fine D, Shoemaker R, Gazdar A, Mayo J, Fodstad Ø, Boyd M, Abbott B and Donovan P: Metastasis models for human tumors in athymic mice: Useful models for drug development. Cancer Detect and Prev 1987 Suppl I: 291-299
33. Fidler IJ: Rationale and methods for the use of nude mice to study the biology and therapy of human cancer metastasis. Cancer Metastasis Rev 1986 (5): 24-29
34. Houchens DP, Ovejera AA, Riblet SM and Slagel DE: Human brain tumor xenografts in nude mice as a chemotherapy model. Eur J Cancer Clin Oncol 1983 (19): 799-805
35. Silverberg E: Cancer Statistics. CA 1985 (35): 19-56
36. Leob LA, Ernster VL and Warner KE: Smoking and lung cancer: An overview. Cancer Res 1984 (44): 5490-5958
37. Fodstad Ø, Aamdal S, Tveit KM and Pihl A: Lung colony formation in adult nude mice upon intravenous injection of cells from a human melanoma xenograft. Proc Amer Assoc Cancer Res 1982 (23): 881
38. Shoemaker R, Wolpert-DeFilippes M, Mayo J, Abbott B, Plowman J, Venditti J, Boyd M, Fodstad Ø, Ovejera A, Donovan P, Griswold D and Dykes D: Experimental chemotherapy studies of a human melanoma in nude mice using a survival endpoint. Proc Amer Assoc Cancer Res 1985 (26): 330
39. Hamilton TC, Young RC, Louie KG, Behrens BC, McKoy WM, Grotzinger KR, Ozols RF: Characterization of a xenograft model of human ovarian carcinoma which produces ascites and intraabdominal carcinomatosis in mice. Cancer Res 1984 (44): 5286-5290
40. Giovanella BC, Yim SO, Morgan AC, Stehlin JS and Williams LJ: Metastases of human melanomas transplanted in "nude" mice. J Natl Cancer Inst 1973 (50): 1051-1053
41. Kuroki M: A new human osteosarcoma strain (Os-Beppu) with a high rate of pulmonary metastasis in nude mice. Jpn J Clin Oncol 1984 (14): 49-56
42. Kerbel RS, Man MS, Dexter D: A model of human cancer metastasis: Extensive spontaneous and artificial metastasis of a human pigmented melanoma and derived variant sublines in nude mice. J Natl Cancer Inst 1984 (72): 93-108
43. Kozlowski JM, Fidler IJ, Campbell D, Xu XL, Kaighn ME and Hart IR: Metastatic behavior of human tumor cell lines grown in the nude mouse. Cancer Res 1984 (44): 3522-3529

VIII. Importance of Drug Resistance in Screening

Mechanisms of Drug Resistance in Screening*

B. W. Fox

Paterson Institute for Cancer Research, Wilmslow Road, Manchester M20 9BX, United Kingdom

General Nature of Resistance

Resistance is a term which should only properly be applied to a single cell. It is the description of the sensitivity which a single cell exhibits towards a drug, applying particularly to the case where the sensitivity is lower than that of the parent cell with which the cell is connected by genetic descent or by comparison with a cell of similar phenotype.

The property of the whole tumour mass is a composite property not only of the tumour cells present but also of the non-tumour cells present such as lymphocytes, macrophages and blood cells as well as those involved with the architecture of the tumour mass itself. The increased insensitivity of this type of tissue should perhaps be more properly referred to as an increased tolerance of the tumour tissue. This suggests that a subpopulation change has occurred under pressure of the drug, to create a tumour which contains proportionally more resistant cells and will, on average, be less sensitive as a whole to the challenging drug.

The underlying mechanisms which create the reduced sensitivity in each case could be different fundamentally and without a knowledge of the details at the molecular levels could lead to erroneous interpretations of the origin of the changed sensitivity towards the drug.

Resistance, which will include tolerance here, can be studied at many different levels as is illustrated in Table 1. There are many examples of increases of biochemical activity which lead to resistance as well as decreases (Table2). Most of these biochemical changes have been identified as a single mechanism in tumours or in cell populations. There are few, if any, reports which have unequivocally identified a number of different lesions within the same tumour mass or a tumour population following development of tolerance within the total population as a result of pressure from a single drug. It is also still not clear if multidrug resistance may not be a subcellular mixture of different basic mechanisms, one mechanism per cell, which overall within the total tumour population appears to result in a multidrug resistance. It would be necessary to prove that several clones of cells derived from single cells from within the tolerant tumour population had genetic and phenotypic identity in terms of their relative resistance to the drugs involved.

Table 1. Levels of study of resistance

1. Tumour subpopulation
2. Cell surface/membrane
3. Cytoplasm
4. Gene sequence
5. Nucleosome
6. DNA & other target sites

Table 2. Mechanisms of resistance involving changes in biochemical activities

Increases in:
1. Enzyme activity normally blocked by drug
2. Repair of damaged DNA
3. Concentration of protective groups
4. Rate of drug breakdown
5. Genetic adaptation (e.g. Gene amplification)
6. Drug efflux
7. Metabolic cooperation, etc.
Decreases in:
1. Enzyme activity that converts drug to active form
2. Eermeability of the drug
3. Enzyme sensitivity towards drug

* The aims of this presentation are to
  - Briefly review the mechanisms of resistance to a number of key agents
  - Identify those forms of resistance associated with xenografts
  - Point to techniques that may be more applicable to studying aspects of resistance in xenografts

Some Examples of Resistance Mechanisms

Cytosine arabinoside. The mechanism of action of this drug is illustrated in Table 3. The prime cytotoxic action of the agent would now not appear to be directly related to its uptake into DNA (1). Resistance is due to at least 3 different mechanisms. Thymidine and cytosine arabinoside have been shown to be synergistic (2) and by the use of 2 resistant lines, the ara C-1-6 T lymphoma which is deficient in dCMP deaminase and the dGuo-200-1 T lymphoma which possesses a mutation in M1 subunit of the ribonucleotide diphosphate reductase gene, it was shown that both were resistant due to an increase in the dCTP levels but that TdR reverses ara C-1-6 but not dGuo-200-1 resistance.

Vinca Alkaloids. The mode of action of this group of agents and 2 features associated with the development of resistance are shown in Table 4. A resistant murine tumour (SEWA) with resistance towards vincristine 30–50 times normal was described by Dahllof et al. (3). The degree of resistance was associated with an increase in abnormal proteins.

Anthracyclines. These agents been extensively studied at different levels of biochemical study, and several changes are known to occur following action of these agents (Table 5). The known mechanisms of resistance include a characteristic increase in the efflux kinetics of the drug for the resistant cells and this mechanism has been exploited, notably by Tsuruo and his colleagues (4), to reestablish resistance by blocking this efflux mechanism. Many calcium channel blocking agents are known to be able to effect this reversal, but it is also known that the reversal effect is not directly or simply related to the calcium channel blocking activity (5).

The uptake of some of these anthracyclines can be followed by fluorescence under UV light of suitable exciting wavelength, and this fact has been used in the flow cytometric study of populations of sensitive and resistant cells (6). This allows for a quantitative assessment on a cell by cell basis of the relative uptakes of the drug within a tumour cell population. This type of study will be necessary for complex tissue structures such as the human tumour xenograft, either in cell culture or more particularly as a solid tumour transplant in an immune deprived mouse, where the total structure of the tumour will be continually changing as the human tumour becomes accommodated to its murine host.

Stepwise increases in the exposure of cells to Doxorubicin or to daunorubicin often result in a form of resistance which is cross-resistant to several other anticancer drugs, often natural products, e.g. vinca alkaloids, actinomycin, colchicine. This has also been shown to be related in some way to the presence of a 160–180 K glycoprotein (7) and in other work to the loss of a normally occurring 100 K glycoprotein (8). A human sarcoma cell line (MES-SA) has been described by Sikic et al., which is 100 times more resistant to doxorubicin, 160 times to daunorubicin and 105 times towards vinblastine than the parent line. In all cases there was an increased efflux of the drugs, but the differences in resistance cannot be related to the differences in drug accumulation alone. It was shown that there was a marked decrease in the membrane fluidity in the resistant cells but no abnormal glycoproteins were observed.

Table 3. Mechanisms of cytosine arabinoside

Action
Phosphorylated to triphosphate by deoxycytidine kinase
Inhibits DNA polymerase II & III not I
Enters DNA & RNA
Enters and damages histone synthesizing in RNA
Resistance
Decrease in deoxycytidine kinase activity
Increase in deoxycytidine-deaminate activity but always accompanied by 1
No evidence for DNA polymerase changes

Table 4. Mechanisms of Vinca Alkaloids

Action
Causes dissolution of mitotic spindle
Arrest of mitosis - Vinblastine not active - reversible
Vincristine less reactive - reversible
Chromosome change
Microtubule crystals containing drug formed
Degeneration of myelin sheath caused
Resistance
Increase in surface glycoproteins
Modified membrane transfer

Table 5. Mechanisms of Anthracyclines

Action
Intercalates with DNA
(a) Inhibits DNA & RNA synthesis
(b) Extensive, irreversible chromosomal damage
(c) DNA breaks
Greatest effect on late S phase - heterochromatin duplication
Forms peroxy radical ions in vivo, damages tissues especially lacking superoxide dismutase e.g., heart muscle
Resistance
Increase in efflux from the cell
Decreased influx (rare)
Cross-resistant to vinca alkaloids usual

Table 6. Mechanisms of Methotrexate

Action
Inhibition of dehydrofolate reductase (DHFR)
Resistance
- Cells have a considerably increased DHFR activity due to gene amplification
- Decreased permeability of MTX
- Change in isoenzyme concentrations

A further example of resistance to these agents in human cell lines were reported by Bhalla et al. (10), who described a promyelocytic leukemia cell line (HL-60/ar) which was isolated in vitro by subculturing methods in increasing levels of doxorubicin and was subsequently found to be fifty-fold more resistant to the challenging drug than the parent cell line (HL-60).

These cells were cross-resistant to daunorubicin, vincristine, vinblastine, actinomycin D but retained sensitivity to methotrexate and ara-C. Two prominant glycoproteins were shown to be present, of molecular weight 160 +/− 10 K and 110 +/− 10 K. Another report of Ahmed (11) using the same cell line, described a form of HL-60 resistant to daunorubicin which was also cross-resistant to doxorubicin, VP16 and to mitoxantrone, and which was characterised as a transport resistant cell.

Methotrexate. The mode of action of this agent has been very well studied as also its mechanism of resistance (Table 6). Considerable degrees of resistance can be achieved, sometimes showing a small degree of collateral resistance to the vinca alkaloids (12). The massive increase in dihydrofolate reductase activity is achieved is by gene amplification, up to almost 200 times, resulting in visible chromosome changes, e. g. homogenously staining regions (HSRs). In human KB cells such changes are often associated with chromosome 10q.

Tanaka et al. (13) described a human gestational choriocarcinoma cell line, resistant to methotrexate to a low level (5 times parent cell sensitivity). In this case, the mechanism of resistance could be related to the lower levels of membrane transport of the drug.

In another example, Mini et al. (14) described a human T lymphoblastic cell line, CCRF-CEM/R1, which was resistant to methotrexate due to an increase in the dihydrofolate reductase activity to about 70-fold. This line was subjected to further methotrexate treatment and they were able to show that there was a further increase in the DHFR activity accompanied by a decreased methotrexate transport. The final sensitivity of this cell line was some 350-fold lower that the original sensitive cell line.

Alkylating Agents. There is a tendency for clinicians to lump together all alkylating agents under the same flag from the clinical point of view. There could be some justification for this when comparing agents with a nitrogen mustard active group and a variety of "carrier" groupings where quantitative differences are seen but fewer qualitative characteristics.

However, it has become clear in the earlier work of Franz Schabel (17) that, within single cell systems, cross resistance does not necessarily imply that similar mechanisms are involved. Thus it is conceivable that a combination of resistance mechanisms characteristic of a particular drug may in fact have been produced. Resistance does seem to be highly drug specific (18).

Resistance to alkylating agents can result in cell membrane permeability changes, free sulphydryl level changes, differences in repair capacity and to clear differences in chromosome sensitivity to drugs. It appears that in the case of possible cross-linking agents, cell killing may be related very closely to the level of unrepaired DNA interstrand cross-links but it is difficult to prove this, since the most sensitive technology for the detection of such lesions is probably several orders of magnitude too insensitive to detect the very few lesions needed to create cell death by this mechanism.

There are many references to the development of resistance to alkylating agents in human cell lines but very few to resistant xenografts. Thomas et al. (19) examined a human colon carcinoma xenograft resistant to methyl CCNU and concluded that resistance in this case is related to the level of measurable cross-links.

Flow Cytometry

The development of higher levels of DHFR enzyme in methotrexate resistant cell lines is an ideal subject for the use of flow cytometry techniques of analysis. A fluorescent complex of fluorescein bridged to methotrexate through a putrescine chain has been synthesised (14) and shown to be effective in distinguishing those resistant cells with high levels of the enzyme from the low enzyme series of sensitive cells. In an artifically mixed population of sensitive and resistant cells it has been possible to distinguish the 2 populations with this probe and also to observe a preferential removal of the resistant line by means of an appropriate concentration of a vinca alkaloid. Other fluorescent analogues have

been described by Kaufmann et al. (15) and a dansyl derivative by Susten et al. (16).
In xenograft systems, where the tumour mass is even more complex than in the other murine tumour systems, where presumably the tumour and non-tumour cells of the tissue mass are changing as host acclimatisation is taking place, it will be necessary to employ techniques designed to study cells at the individual cell level and flow cytometry offers the best means of quantitating this. A combination of this technique and histopathological examination using more advanced probes would be a possible route to understand the complex parameters which underly resistance in xenografts.

Conclusions

1. There are now many known mechanisms by which tumour cells protect themselves against the damaging action of anticancer agents.
2. Human cell lines in culture have been shown to acquire resistance by increasing exposure to anticancer drugs. There are very few instances only where resistance in xenografts in situ have been studied mechanistically.
3. It is still too early to assess whether tumour xenografts show the same spectrum of resistance mechanisms as murine transplanted tumours.

References

1. Crowther PJ, Cooper IA, Woodcock DM: Cancer Res 1985 (45): 4291
2. Cohen A, Willman B: Cancer Chemotherap Pharmacol 1985 (14): 70
3. Dahllof R, Martinsson T, Levan G: Exp Cell Res 1984 (152): 415
4. Tsuruo T, Kavabata H, Nagumo N et al.: Cancer Chemotherap Pharmacol 1985 (15): 16
5. McGown AT, Fox BW (submitted for publication)
6. McGown AT, Ward TH, Fox BW: Cancer Chemotherap Pharmacol 1983 (11): 113
7. Juliana RL, Ling RV: Biochim Biophys Acta 1976 (455): 152
8. Biedler JL, Chang T-d, Meyers MB, Peterson RHF, Spengler BA: Cancer Treatment Rep 1983 (67): 859
9. Sikic BI, Etiz BB, Haeker WG, Anderson RL, Hahn GM: Proc Ann Meet Amer Assoc Canc Res 1985 (26): 34
10. Bhalla K, Hindenberg A, Taub RN, Grant S: Cancer Res 1985 (45): 3657
11. Ahmed NK: Proc A A C R 1985 (26): 337
12. Poppitt DG, McGown A, Fox BW: Cancer Chemotherap Pharmacol 1984 (13): 54
13. Tanaka K: Nippon Senka Fujinka Gakkai Zasshi 1984 (36): 2129
14. Mini E, Moroson BA, Franco CT, Bertino JR: Cancer Res 1985 (45): 325
15. Kaufmann RJ, Bertino JR, Schimke RT: J Biol Chem 1979 (253): 5852
16. Susten SS, Kempton RJ, Black AM, Freishan JG: Biochem Pharmacol 1980 (33): 1957
17. Schabel FM, Skipper HE, Trader MW, Laster WR, Griswold DP, Corbett TH: Cancer Treatment Rep 1983 (45): 3460
18. Bodell WJ, Gerosa M, Aiden T, Berger MS, Rosenblum ML: Cancer Res 1985 (45): 3460
19. Thomas CB, Osieka R, Kohn KW: Cancer Res 1978 (38): 2448

Circumvention of Inherent Resistance with Verapamil in a Human Tumour Xenograft

S. MERRY*, D. CUNNINGHAM**, E. R. COURTNEY, T. HAMILTON, S. B. KAYE, and R. I. FRESHNEY

Department of Medical Oncology, University of Glasgow, 1, Horselethill Road, Glasgow G12 9LX, Scotland, United Kingdom

Abstract

The calcium antagonist verapamil has been used to overcome multidrug resistance in a number of animal and human tumour models. *In vivo* data derived from human solid tumours is, however, limited. We have investigated the ability of verapamil to increase sensitivity to etoposide (VP16), adriamycin (ADR) and vincristine (VC) in the human non-small cell lung cancer tumour line WIL. *In vivo* cytotoxicity was measured using subcutaneous xenografts in MF1-nu/01a (Olac Ltd) nude mice with caliper measurements to determine tumour volume doubling time following a single i.p. dose of drug. To determine *in vitro* cytotoxicity exponentially growing cells were exposed to cytotoxic drugs with- or without verapamil for 72 hours and, after a recovery period of 120 hours, viability was assessed as incorporation of tritiated leucine. Plasma pharmacokinetics of verapamil (25 μg/g i.p.) were determined by HPLC. *In vivo* dose-response curves showed WIL to be insensitive to VP16, ADR and VC with increases in tumour volume doubling time of 8%, 9% and 17%, respectively, at the approximate LD_{10} doses (i.e. 40 μg/g, 10 μg/g and 3 μg/g). The addition of 25 μg/g i.p. verapamil (approximate LD_{10} dose) to this treatment, however, produced increases of 46% ($p<0.01$), 28% ($p<0.1$) and 4% relative to treatment with cytotoxic drug alone. These results show a correlation with *in vitro* results in which 6.6 μM verapamil reduced ID_{50} by 13-fold, 8-fold and 3-fold, respectively. Plasma pharmacokinetics of verapamil showed a peak concentration of 1.4–1.6 μM and *in vitro* cytotoxicity assays using 2.2 μM verapamil showed an increase in sensitivity for VP16 (4-fold), but not for ADR of VC. These data indicate that effects at the cellular level may be responsible for the ability of verapamil to increase the *in vivo* chemosensitivity of a human lung cancer tumour line, and furthermore suggest a potential clinical role for verapamil in human solid tumours.

* To whom correspondence should be addressed

** Present address: Department of Medical Oncology, St. George's Hospital, Blackshaw Road, London SW17 0QT, United Kingdom

Introduction

It is a common observation *in vitro* that the calcium antagonist verapamil is able to circumvent resistance to the cytotoxic drugs adriamycin (ADR), vincristine (VC) and VP16 (1, 2). This observation, for VC, has been extended to mice bearing the murine leukaemia P388 as an ascites (3) and the murine colon adenocarcinoma C26 implanted i.p. as a suspension of tumour fragments (4). In both these cases, however, the animals were treated i.p. with vincristine plus verapamil beginning one day after tumour implantation. The effect of verapamil on the treatment of established tumours is investigated in this study using nude mice bearing subcutaneous human lung adenocarcinoma xenografts of the tumour line WIL. The use of this model enables human tumours growing *in vivo*.

Materials and Methods

Drugs

The drugs used in this study were adriamycin (Farmitalia Carlo Erba Ltd, Barnet, Herts, U.K.), verapamil (trade name "Cordilox I.V.", Abbott Laboratories Ltd, Queenborough, Kent, U.K.), vincristine sulphate (trade name "Oncovin", Eli Lilly and Co Ltd, Basingstoke, U.K.) and VP16-213 (etoposide, trade name "Vepesid", Bristol Myers Pharmaceuticals, Slough, U.K.). They were solubilised according to the manufacturers' instructions and then stored at −20 °C. They were further diluted in saline (for experiments *in vivo*) or tissue culture medium (for experiments *in vitro*) as required immediately prior to use.

Animals

MFI-nu/Ola nude (6–12 weeks old) and (C57 BL/6 × DBA/2)F hybrid immunocompetent mice (aged 14–20 weeks) of either sex were used in these studies with conditions of lighting and feeding kept constant throughout the experiments. The nude mice were bred in the Department of Medical Oncology, Glasgow University and the immunocompetent mice were bred in the Royal Infirmary, Glasgow.

Tumour Model

The human lung adenocarcinoma cell line WIL was obtained from Dr. M. O'Hare (Ludwig Institute for Cancer Research, Sutton, Surrey, U.K.) as a cell line growing *in vitro*. It was established as a xenograft by subcutaneous injection of a cell suspension and then serially passaged *in vivo* using approximately 20 mm^3 tumour fragments.

WIL was also maintained *in vitro* as a monolayer cell line. The growth medium consisted of Ham's F10/Dulbecco's Modified Eagles Medium (50:50, v/v) supplemented with 10% foetal calf serum and with a gas phase of 2% CO_2. In some cases the medium was also supplemented with 50 μg/ml gentamycin.

Drug Sensitivity Measurements In Vivo

Subcutaneous xenografts. Tumour volume (v) was calculated using the formula $v = (\pi/6) \times d^3$ where d is the diameter (average of 2 determinations) of the tumour measured using calipers. This formula (5) assumes the tumour to be ellipsoid in shape. In preliminary experiments a good correlation ($r = 0.96$; $n = 66$) was obtained between measured volume and weight of excised tumour. This correlation was obtained using 20 untreated tumours and 46 tumours treated with cytotoxic drugs.

In drug sensitivity experiments groups of mice (5–10 animals) with tumours of 0.5–1.5 cm^3 were treated i.p. with the approximate LD_{10} dose of cytotoxic drug (10 μg/g ADR, 3 μg/g VC or 40 μg/g VP16). The animals were also treated with 25 μg/g verapamil or an equivalent volume of saline as a second i.p. injection within 1 minute of the administration of the cytotoxic drug. This dose of verapamil also approximated the LD_{10} dose.

Tumour size was determined by twice-weekly caliper measurements and the time required for individual tumours to undergo 2 volume doublings (i.e. to increase to 4 times their original volume) was determined. Tumour volume doubling time was expressed as one half of this value.

The sub-renal capsule assay. The method has been described previously (6) and was based on that of Bogden et al. (7). Basically 1 mm^3 tumour pieces were placed under the kidney capsule of immunocompetent mice on day 0. On days 1, 3 and 5, groups of 6 mice were treated with 25 μg/g verapamil i.p. or saline, followed (after 30 minutes) by 40 μg/g VP16 subcutaneously. These total doses of verapamil and VP16 represent 3 times those used in the experiments with subcutaneous xenograft, although the route of administration of VP16 was different. Control mice were treated with 2 saline injections. Tumour growth was determined on day 6 from increase in 2 perpendicular diameters of the tumour using a stereoscopic microscope with a micrometer scale.

Verapamil Pharmacokinetics

At time 0 non-tumour bearing nude mice were treated with 25 μg/g verapamil i.p. Pairs of mice were killed at each time point (see Fig. 1) by exsanguination (from the vena cava) under ether anaesthesia. The blood was collected in heparin-containing tubes, centrifuged, and plasma samples were stored at - 20 °C prior to analysis by Dr. M.D. Brodie (Department of Medicine, Western Infirmary, Glasgow) using an HPLC technique to determine levels of verapamil and its metabolite norverapamil.

Drug Sensitivity Measurements In Vitro

The assay used has been described previously (8). Briefly, 10^3 cells in 0.2 ml of growth medium were seeded into a 96-well tissue culture plate. The cells were allowed to enter exponential growth (96 hours with a medium change after 72 hours) and were then exposed to cytotoxic drugs for 72 hours (with drug and medium replacement at 24 and 48 hours) followed by a recovery period of 120 hours (with 2 changes of medium) in the absence of drug. In experiments to determine the effect of verapamil on cytotoxic drug sensitivity, verapamil was added to the culture medium only during the period of cytotoxic drug exposure.

At the end of the assay viability was assessed as incorporation of radiolabelled leucine into the trichloroacetic acid insoluble material derived from the cells. From dose-response curves the ID_{50} (concentration of cytotoxic drug which caused a 50% reduction of viability) was determined.

Results

Table 1. Effect of verapamil on the sensitivity of subcutaneous WIL xenografts to cytotoxic drugs

Treatment	ADR (10 µg/g)	VC (3 µg/g)	VP16 (40 µg/g)
1) saline only	6.15 ± 0.80[a]	7.37 ± 1.02	5.81 ± 1.30
2) 25 µg/g verapamil only	5.28 ± 0.27	7.05 ± 1.05	6.88 ± 0.92
3) cytotoxic drug only	6.70 ± 0.75	8.65 ± 1.06	6.26 ± 0.32
4) cytotoxic drug plus 25 µg/g verapamil	8.60 ± 0.70 (p < 0.1)	8.98 ± 2.10	9.16 ± 0.79 (p < 0.01)

[a] Tumour doubling time (days): mean ± sem.

Table 1 shows the results of our drug sensitivity experiments *in vivo*. In all experiments, the dose of verapamil used was not found to be cytotoxic. For VP16, a statistically significant (Student's t test, $p < 0.01$) increase in tumour volume doubling time was produced when verapamil was combined with the cytotoxic drug. An increase was also seen in the case of ADR, although this did not attain statistical significance ($p < 0.1$). For VC no increase in tumour volume doubling time was noted. With the exception of vincristine, the combination of verapamil with cytotoxic drugs did not cause a significant increase in mortality. We found 0% mortality for verapamil plus VP16, 0% mortality for verapamil plus ADR and 55% mortality for verapamil plus VC. These observations must, however, be regarded as preliminary since they are based on small numbers of animals.

Fig. 1 shows the results of the subrenal capsule assay experiments. In this model the combination of VP16 and verapamil also resulted in a statistically significant reduction in tumour growth compared to treatment with VP16 alone.

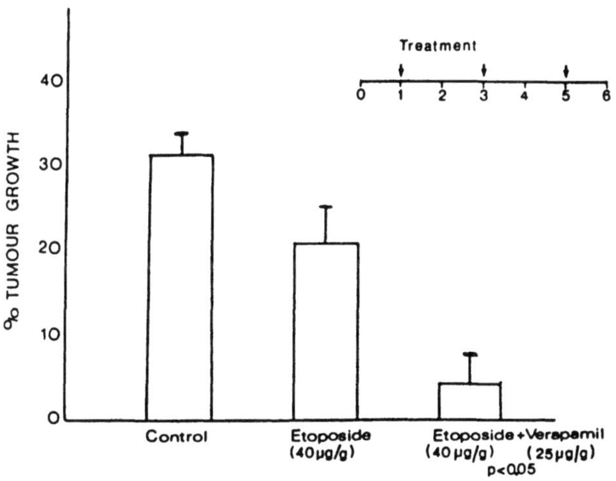

Fig. 1. Effect of verapamil on the sensitivity of WIL to VP16 in the 6-day subrenal capsule assay. The histograms present mean +/− 1 standard error

The plasma pharmacokinetics of verapamil and its metabolite norverapamil in nude mice are shown in Fig. 2. Preliminary data *in vitro* (not shown) has indicated that norverapamil has similar activity to verapamil in potentiating cytotoxic drug efficacy. In these experiments verapamil was administered at the approximate LD_{10} dose. Peak plasma levels of verapamil of approximately 1.5 µM were obtained 20-60 minutes after injection. At these times the levels of norverapamil were approximately 0.5 µM. The plasma concentration of norverapamil rose to 0.7 µM 2 hours after injection.

The cytotoxicity data obtained *in vitro* is shown in Table 2. In all cases the range of variation between replicates is less than 42% of their mean value. A non-cytotoxic concentration of verapamil (6.6 µM) decreased ID_{50} by 12.5 fold for VP16, 8.4 fold for ADR and 3.3 fold for VC. When 2.2 µM verapamil was used (a concentration which approximates the peak level achieved in nude mice, Fig. 3) a decrease in ID_{50} for VP16 of 3.9 fold was noted.

Discussion

We have determined the effect of verapamil on the sensitivity of the human lung adenocarcinoma tumour line WIL to the cytotoxic drugs ADR, VC and VP16 both *in vivo* and *in vitro*. As resistance in this tumour line was not maintained by routine exposure to cytotoxic drugs in the laboratory, our data show the ability of verapamil to overcome inherent resistance in this model. Verapamil increased the sensitivity of WIL to VP16 using a) subcutaneously-growing tumours in nude mice (Table 1) and b) the subrenal capsule assay in immunocompetent mice (Fig. 1). Verapamil has been previously shown to circumvent resistance to VP16 *in vitro* using Ehrlich ascites tumour cells (9) and a human T cell lymphatic leukaemia cell line (10). Verapamil has also been shown to potentiate the antitumour activity of VP16 *in vivo* using mice bearing murine ascites tumours (10). To our knowledge verapamil has not previously been shown to circumvent inherent drug resistance in a human tumour line *in vivo*.

The mechanism of action of verapamil in our model remains unclear, but (for ADR, VC and VP16) increases in sensitivity *in vivo* (Table 1) are paralleled by increases in sensitivity *in vitro* (Table 2), i.e. effect on VP16 > ADR > VC. This suggests that effects at the cellular level (rather than gross pharma-

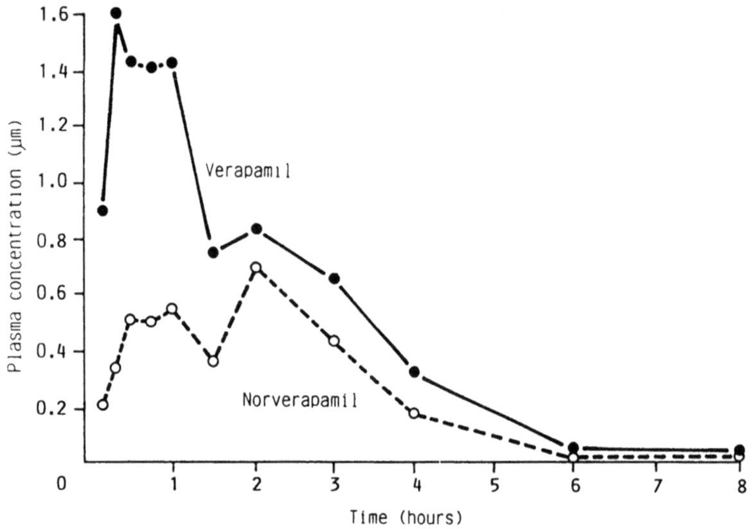

Fig. 2. Plasma pharmacokinetics of verapamil and its major metabolite norverapamil in the nude mouse. Each point is the mean of duplicate determinations. Mice were treated with 25 μg/g verapamil i.p. at time 0

Table 2. Effect of verapamil on the sensitivity of WIL *in vitro* to cytotoxic drugs

VERAPAMIL CONCENTRATION ($\times 10^{-6}$M)	ADR ($\times 10^{-10}$M)	VC ($\times 10^{-9}$M)	VP16 ($\times 10^{-9}$M)
0	168[a]	136	700-830
2.2	168-186	96-148	192-196
6.6	20	34- 52	55- 70

[a] Drug ID_{50} (range of replicate determinations).

cological changes) may be responsible for the observed action of verapamil *in vivo.* In a number of animal and some human models increases in sensitivity produced by verapamil have been shown to be associated with increased intracellular cytotoxic drug levels, possibly due to reduced cellular drug efflux (2), and verapamil may have a similar mechanism of action in our model.

We would stress that our studies were carried out using a tumour line with inherent drug resistance. In a study of arabinosyl-5-azacytosine (Ara-AC) Ahluivalia et al. (11) have shown differences between the mechanisms of resistance of a P388 murine leukaemia with induced resistance produced by continuous culture in the presence of Ara-AC and a murine colon carcinoma with inherent resistance to this drug. Our results, however, indicate that the mechanisms of resistance of WIL to ADR, VC and VP16 may be similar to those of models in which resistance was derived by repeated drug exposure in the laboratory as both can be circumvented with verapamil.

Acknowledgement

We thank the Cancer Research Campaign for financial support and Mrs Margaret Jenkins for typing the manuscript.

References

1. Chabner BA, Clendeninn NJ and Curt GA: Symposium on cellular resistance to anticancer drugs: Introduction. Cancer Treat Rep 1983 (67): 855-857
2. Kaye S and Merry S: Tumour cell resistance to anthracyclines - a review. Cancer Chemother Pharmacol 1985 (14): 96-103
3. Tsuruo T, Iida H, Tsukagoshi S and Sakurai Y: Overcoming vincristine resistance in P388 leukaemia in vivo and in vitro through enhanced cytotoxicity of vincristine and vinblastine by verapamil. Cancer Res 1986 (41): 1967-1972
4. Tsuruo T, Iida H, Naganuma K, Tsukagoshi S and Sakurai Y: Promotion by verapamil of vincristine responsiveness in tumour cell lines inherently resistant to the drug. Cancer Res 1983 (43): 808-813
5. Shorthouse AJ, Jones JM, Steel GG and Peckham MJ: Experimental combination and single-agent chemotherapy in human lung-tumour xenografts. Br J Cancer 1982 (46): 35-44
6. Cunningham D, Jack A, McMurdo DFS, Soukop M, McArdle CS, Carter DC and Kaye SB: The 6 day subrenal capsule assay is of no value with primary surgical explants from gastric cancer. Br J Cancer 1986 (54): 519-523
7. Bogden AE, Kelton DE, Cobb WR and Esber JH: A rapid screening method for testing chemotherapeutic agents against human tumour xenografts. In: Houchens DP and Orejera AA (eds) Proceedings of the symposium on the use of athymic (nude) mice in cancer research. New York: Gustav Fischer Inc 1978: 231-250
8. Merry S, Kaye SB and Freshney RI: Cross-resistance to cytotoxic drugs in human glioma cell lines in culture. Br J Cancer 1984 (50): 831-835
9. Seeber S, Osieka R, Schmidt CG, Achterrath W and Crooke ST: In vivo resistance towards anthracyclines, etoposide and cis-diamminedichloroplatinum (II). Cancer Res 1982 (42): 4719-4725
10. Slater LM, Murray SL, Wetzer MW, Sweet P and Stupecky M: Verapamil potentiation of VP16-213 in acute lymphatic leukaemia and reversal of pleiotropic drug resistance. Cancer Chemother Pharmacol 1986 (16): 50-54
11. Ahluiwalia GS, Cohen MB, Kang CJ, Arnold ST, McMahon JB, Dalal M, Wilson YA, Cooney DA, Balzarini J and Johns DG: Arabinosyl-5-azacytosine: mechanisms of native and acquired resistance. Cancer Res 1986 (46): 4479-4485

IX. Proferred Papers at the E.S.O. Seminar

Clonogenic Assay Screening Study Group (CASSG): Contribution to an EORTC Drug Screening Programme

M. S. Aapro

For the CASSG, University Hospital, 1211 Geneva 4, Switzerland

It is usual that drug screening programmes are changed after limitations of previous setups are realized. It is essential to weigh the advantages and limitations of various techniques in the overall design of a new drug development strategy. The laboratories of the CASSG, which can use several techniques, are currently evaluating the reproducibility of drug testing in colony forming cell assays (Rozencweig M, Proc. Am. Assoc. Cancer Res 1983 (24): 313). They have shown that resistance and sensitivity are reliably predicted. (Uitendaal M, in preparation). A continuous quality control study to check reproducibility of data is currently being run (Eliason JF, chairman).

As primary (non-specific antitumor) screening cannot be performed on fresh tumor samples (Shoemaker RH et al., Cancer Res 1985 (45): 2145), the CASSG has developed a proposal to evaluate drugs against tumor cell lines in clonogenic assays. Studies will be done in several laboratories and some will be repeated in at least 2 independent laboratories. Cell lines might include brain, breast, non-small cell and small cell lung, colon, melanoma and ovary. They will also be characterized on the basis of sensitivity to most common drugs. Lines with different profiles will be used. If drugs are chosen for clinical development on the basis of these and other tests, they will be tested against human bone marrows and fresh tumor cells. This will define hemopoietic toxicity and target the most sensitive tumor types. It will confirm indications obtained from prior dose-response testing of compounds of interest against cell lines. The contribution of the CASSG to an EORTC drug screening programme will not reproduce the strategy of the NCI (Shoemaker RH, Cancer Treat Rep 1986 (70): 9) as cell lines and techniques used will not be the same. Coordination between the NCI and EORTC will however allow some cross-screening and validation of each other's results. The exact breadth of the CASSG contribution will depend on priorities discussed with the EORTC New Drug Development Office.

Technical developments may require adaptation of the group's proposal.

Polyamine Uptake by Malignant Brain Tumour Xenografts in Nude Mice

F. DARCEL[1], M. CHATEL[1], V. QUEMENER[2], and J. PH. MOULINOUX[2]

[1] Laboratoire de Neuropathologie C. H. U. Rennes, France
[2] Laboratoire d'Histologie, C. H. U. Rennes, France

Polyamines (putrescine, spermidine and spermine) are aliphatic polycations which have been shown to play a major role in cell proliferation and differentiation. As far as brain tumours are concerned, the following data were found:
In medulloblastomas, an increase of the C. S. F levels of putrescine and spermine during phases of clinical deterioration suggested the use of these compounds as biological criteria of recurrence. (Marton et al.); in malignant astrocytomas, an elevation of ornithine and S-Adenosyl methionine decarboxylases, key enzymes of polyamine synthesis, has been shown to be correlated with the grade of malignancy (Scalabrino et al.); lastly, erythrocyte polyamine levels in patients harbouring malignant gliomas are statistically correlated to phases of proliferation and may be used for the therapeutic monitoring of adjuvant therapy (Moulinoux et al.).
Having obtained these data, we planned to explore the polyamine metabolism in patients by means of positron emission tomography (P. E. T.).
As a first step, we decided to evaluate the polyamine uptake by glioma heterografts in nude mice after intraperitoneal injection of labelled putrescine, spermidine and spermine.
The polyamine metabolism in a cell is the result of intrinsic synthesis on the one hand and extracellular uptake on the other. The question to be answered in this protocol was: is this intracellular pathway correlated with the tumour proliferative index and, if so, is the tumour versus normal tissue uptake ratio important enough to be used in a P. E. T. study?
Series of human glioma grafted nude mice were intraperitoneally injected with I uc of ^{14}C putrescine, ^{14}C spermidine and ^{14}C spermine. The injection schedule was planned taking into account the size of the tumours.

Two conclusions were reached:
- Only putrescine uptake was significantly increased in the tumour tissue versus normal brain, while the ratio was not significant for spermidine and spermine.
- The putrescine uptake by mg of tumour tissue is dependent upon tumour size (decreasing progressivly as the tumour volume increased).

These preliminary findings are in agreement with experimental data obtained in intracerebral grafted rat glioma (Volkow et al.), suggesting that positron emitting putrescine could be a useful probe for exploring human glial tumour proliferation *in vivo.*

The Activity for a Novel Niteosourea (TCNU) in Human Lung Cancer Xenografts

R. J. FERGUSSON, L. E. ANDERSON, and J. F. SMYTH

Imperial Cancer Research Fund Medical Oncology Unit, Western General Hospital, Edinburgh, Scotland, United Kingdom

The activity of a novel water-soluble taurine-containing nitrosourea 1-(2-chloroethyl)-3-/2-(dimethylaminosulfonyl)-ethyl/-1-nitrosourea (TCNU) was assessed in 3 human bronchial carcinoma xenografts grown in immunodeficient mice. Animals were prepared by neonatal thymectomy and total body irradiation after a priming dose of cytosine arabinoside. Groups of 6-9 tumours were randomly allocated to either treatment with a single MTD dose of TCNU (20 mg/kg orally) or to a control group. Tumours were measured 3 times per week and the activity of the drug expressed in terms of the specific growth delay (SGD) derived from the median doubling times of treated and control groups. Activity was seen in 2 non-small cell tumours, an adenocarcinoma and a squamous carcinoma (SGDs of 0.67 and 0.96, respectively). In a small cell tumour TCNU was curative at this dose.

Xenografts in Pharmacologically Immunosuppressed Mice as a Model to Test the Chemotherapeutic Sensitivity of Human Tumours

G. L. FLOERSHEIM, A. BIERI, and N. CHIODETTI

Laboratory for Transplantation Biology, Departments of Surgery and Research, Zentrum für Lehre und Forschung, Kantonsspital, Hebelstrasse, 20, 4031 Basel, Switzerland

Growth of human tumor xenografts was achieved in mice treated for 4 days prior to transplantation with an immunosuppressive protocol consisting of 2 combined doses of procarbazine and cyclophosphamide alternating with 2 doses of rabbit anti-mouse thymocyte serum. This model was assessed for its suitability to test preclinically the sensitivity of colorectal carcinomas, bone sarcomas and melanomas against anticancer agents. Besides ionizing radiation, 14 cytotoxic drugs including 5-fluorouracil (5-FU), dimenthylmyleran (DMM), cytosine arabinoside, cyclophosphamide, melphalan, BCNU, mitomycin C, adriamycin, bleomycin, etoposide, vinblastine, cisplatin, procarbazine and DTIC were assayed. Ionizing radiation, 5-FU and DMM were also applied at lethal dosages followed by bone marrow rescue (superdosis therapy). The 4 colon carcinomas responded poorly to most of the agents but one tumor displayed marked sensitivity to BCNU. Lethal doses of radiation, 5-FU and DMM could also show considerable activity. A Ewing sarcoma was highly sensitive to DMM and cyclophosphamide and an osteosarcoma to the latter drug. No strong effects were seen against 2 melanomas. Lethal doses of DMM gave the best regression in a colon carcinoma. In general, the superiority of superdosis therapy for solid human tumors as compared to maximally tolerated doses was demonstrated. Individual carcinomas of the same type displayed a different drug sensitivity. The introduction of this predictive sensitivity assay is likely to allow, by individualizing the tumours, to optimize the therapy for appropriate human cancers.

New Tissue-Specific Metastasis Models in Adult Nude Mice

Ø. FODSTAD, S. AAMDAL, R. H. SHOEMAKER, and A. PIHL

Institute for Cancer Research, The Norwegian Radium Hospital, 0310 Oslo 3, Norway, and DPT, National Cancer Institute, NIH, Bethesda, MD 20205, U.S.A.

We have recently developed several models for studying human tumor metastasis in adult nude mice.

The melanoma line LOX gives, after i.v. injection of 10^6 cells, lung colonies that lead to death of the animals in approximately 25 days. One subline growing s.c. results in the formation of spontaneous lung metastasis, and another subline can be grown i.p. as an ascitic tumor which kills the animals in a dose dependent manner. The FEMX melanoma line gives s.c. and lymph node metastasis but never lung colonies, even though the same fraction of radiolabelled cells is initially retained in the lungs as in the case of the LOX line. The two melanoma lines have been compared with respect to growth rates *in vivo* and *in vitro,* karyotype and expression of membrane antigens.

During recent chemosensitivity studies of a xenografted large cell lung cancer line IMJX, distant s.c. metastasis was observed in a nude mouse in which the primary tumor had vanished in response to therapy. A tumor line established from the metastasis is now being compared with the parent line.

The new models provide opportunities to study the effect to new chemotherapeutic agents on metastasis formation as well as on the growth of established metastases growing at different sites.

Fractionated Irradiation of Human Tumor Xenografts in Athymic Nude Mice

W. HINKELBEIN, A. WUERDINGER, E. WALLBRECHER, and H.H. FIEBIG

Department of Internal Medicine, Division of Hematology/Oncology, Medizinische Universitäts-Klinik, Hugstetter Straße 55, 7800 Freiburg i. Br., Federal Republic of Germany

Two human non-small cell carcinomas of the lung (poorly differentiated squamous cell carcinomas) are implanted into nu/nu NMRI mice. The mice are randomized to an untreated control group or to a therapy group (12 animals in each group) when the median tumor volume is about 1 cm^3. Total doses are 24 and 32 Gy in all fraction schemes (5×2 Gy, 6×3 Gy, 2×6 Gy/week and 3×2 Gy/day, 6 times a week). The treatment machine is a 250 kV x-ray unit with 3,5 mm Al filter. Irradiation is monitored by a PTW-DU 3 dosemeter. The animals are tranquillized by diazepam and chlorpromazine during irradiation and the tumors of 4 mice are fixed in a special treatment cage in order to obtain a homogeneous dose and protect the mice against radiation. Tumor volumes are verified twice a week and related to the beginning of therapy. Besides tumor size the specific regrowth delay is determined. Tumor regression is different depending on the fractionation scheme.

Supported by Grant DFG, Hi 368/1-1

An Animal Model for Human Tumor Xenografts

J. HOOGENHOUT[1], C.R. JERUSALEM[2], J.J.M. VAN GASTEREN[1], and H.B. KAL[3]

Departments of Radiotherapy[1] and Cytohistology[2], St. Radboud Academic Hospital, Nijmegen, and Radiobiological Institute TNO[3], Rijswijk, The Netherlands

Animal models for human tumor xenografts are used for the study of biology and treatment of human cancer. The aim of this study was to develop a model for long-term tumor observations with the possibility of intensive treatment schedules (fractionated irradiation) and host manipulation.

Three-month-old male Wistar rats of the outbred strain Cpb:WU were treated with cyclophosphamide (i.p.) at doses of 10 mg/kg on 4 consecutive days. Two days later, total lymphoid irradiation (TLI) was performed giving a single dose of 9.0 Gy at a dose rate of 0.60 Gy/min. Ten animals could be irradiated simultaneously with the 18 MV photon beam of a linear accelerator.

Four days after TLI tumor xenografts were implanted in subcutaneous tunnels of the immunosuppressed rats. Immunosuppression was maintained with Cyclosporin A (10 mg/kg daily) given orally. The effect of immunosuppression was monitored by lymphocyte counts, determination of serum IgG concentration, PHA and Con A responses, proportions of B cells and histopathological studies of the lymphoid organs. Six to 8 days after TLI the lymphocyte counts, IgG levels, PHA and Con A values were severely decreased. Mitotic activity in the depleted B and T cell areas of the peripheral lymphatic organs was not observed.

At present 4 human tumor lines are serially passaged in the rat xenograft system (1 osteosarcoma, 2 squamous cell carcinomas and 1 neuroblastoma). Xenografts are left to grow up to a period of 200 days, depending on the growth rate of the xenograft. The morphology, take rate and tumor volume doubling time did not change during the first 7 passages. The immunosuppressed rat model is available at a reasonable cost and time. The take rate of tumor xenografts is high (60-96%). The size, durability and life span of the rat makes it suitable for experimental manipulations.
The model is applicable for protracted studies and multipurpose investigations.

Xenografted Human Tumors in Preclinical Drug Design

L. Kopper, P. Nagy, J. Rajnay, and K. Lapis

Institute of Pathology and Experimental Cancer Research, Semmelweis Medical University, Budapest 1085, Hungary

More than 100 human tumors were transplanted into artificially immunosuppressed CBA mice with an overall success rate of 24% (16% colorectal carcinoma, 22% renal cell carcinoma, 12% NHL etc.). These tumor panels were used to evaluate the correspondence between the results obtained with known drugs in clinical practice and in xenograft models. Using the value of "growth delay" for drug ranking the response was considered as positive when over 0.5. Although it is still unknown whether the panel is representative of the tumor population at large, the good agreement between sensitivity of xenografted tumors and clinical experience indicated the usefulness of the model. Several newly developed structures have been tested and some of them produced considerable growth inhibition - alone or in combination (e.g. bifunctional nitrosourea, EUdR + 5-FU for colorectal carcinoma, MSH-fragment nitrosourea for melanoma, DAG for lymphoma etc.).
The subcutaneous (SC) and subrenal capsule (SRG) assay were compared using 2 xenografted lines derived from human tongue carcinoma and its metastasis both in immunocompetent and immunosuppressed mice. Six clinically active agents were applied and the sensitivity of the lines was similar. The SC and SRC assay also showed similar ranking of drugs, i.e., 2 out of 3 most effective agents (CY, 5-FU, VBL) were identical. It is necessary to use immunocompetent mice to check the morphology of the tumors on the day of drug evaluation. These results allow us to conclude that this unique experimental system can provide valuable information in preclinical drug design.

Growth in Nude Mice of a T-Cell Line from a Case of Acute Lymphoblastic Leukaemia: A Model for Experimental Immunotherapy of Human Neoplasia

M. G. MARTINOTTI, R. ARIONE, and G. FORNI

Institute of Microbiology, University of Turin, Turin, Italy

A T-cell line was established *in vitro* from a case of T acute lymphoblastic leukaemia. This line was characterized as far as the surface and chromosomal markers. When these cells were injected subcutaneously in normal or nude Balb/c mice they were rejected. In contrast, a subcutaneous mass was observed when leukaemia cells were injected into nude mice which had been splenectomized 10 days previously and had received total body irradiation (400 rads) 3 days before xenotransplantation. The tumour mass became first evident 2 weeks after transplantation, after which it grew progressively. Histological examination revealed human lymphoblastoid cells with the same surface and chromosomal markers as the original injected cells. When leukaemic cells obtained from the tumour mass were injected back into splenectomized and irradiated nude mice, again a tumour was obtained which became serially transplantable. The growth pattern of this leukaemia was not affected when allogeneic peripheral blood lymphocytes (PBL) were admixed with the leukaemic cells at a 5:1 ratio. In contrast, full inhibition of tumour growth was found when the allogeneic PBLs were previously stimulated in a mixed leukocyte culture against the leukaemic cells. Stimulated allogeneic PBLs displayed a marginal cytotoxicity against leukaemic cells, while substantial amounts of IFN were released. A similar leukaemia inhibition was also found following inoculation of non-stimulated PBLs and daily injections of recombinant IFN (Hoffman La Roche, U.S.A.). The effects of IFN on allogeneic lymphocytes and leukaemic cells are currently under investigation. In any case these findings are of some importance because the serial transplantation of haematopoietic malignant cells is exceedingly difficult.

Supported by the Italian Association of Cancer Research.

5-Fluorouracil (5-FU) + Cis-diamminedichloroplatinum (DDP) Treatment of Human Colorectal Tumor Xenografts

G. PRATESI, M. TORTORETO, and G. PARMIANI

Istituto Nazionale Tumori, Via Venezian 1, 20133 Milano, Italy

5FU + DDP is a new antitumor drug combination of clinical interest. Objective responses have been reported in solid human tumors of different histotypes. Fragments from 2 human colorectal tumors, 1 primary (COMU) and 1 liver metastasis (COFF), were grafted s.c. into both flanks of athymic Swiss nude mice, in order to investigate their sensitivity to this drug combination. The drugs were administered i.v. 3 times weekly; treatment was started when the tumour reached approximately 100-200 mm^3. The doubling time (DT) for both tumors was calculated and the growth delay (GD) assessed for treated (t) as compared to control (c) mice (GD 200% = mean DT_t − mean DT_c/mean DT_c). The results are reported in Table 1. 5FU only slightly reduced the tumor growth rate at 44 mg/kg (3/5 mice died of toxicity), whereas DDP at its optimal dose of 5.9 mg/kg induced a tumor growth delay value of >1 in both tumors. Moreover, when the two drugs were given in combination (24 hr interval), antitumor activity was observed only for the sequence 5FU → DDP, whereas the same doses in the opposite sequence (i.e., DDP → 5FU) were highly toxic. Therefore in this combination therapy, drug sequence strongly affects antitumor activity and toxicity and this factor may explain the contrasting results obtained in different clinical trials on advanced colorectal tumors.

Compound	Dose mg/kg	COMU		COFF	
		DT (days)	GD 200%	DT (days)	GD 200%
Controls	-	9	-	9	-
5FU	44	13	0.44		
DDP	5.9	20	1.22	21	1.30
5FU + DDP	44 + 4	23	1.55	20	1.20
DDP + 5FU	4 + 44			nd (tox)	

Inter and Intra-Tumor Variation in the Response to Chemotherapy of Human Small Cell Carcinoma of the Lung (SCCL) Grown in Nude Mice

M. SPANG-THOMSEN and S. A. ENGELHOLM

University Institute of Pathology, Copenhagen, Denmark

Tumor heterogeneity includes the response to chemotherapy. The inter-tumor response variation of SCCL was investigated by treating 4 xenografted tumors with single dose LD_{10} of melphalan, and the intra-tumor variation to different drugs by treating one of the tumors with single dose LD_{10} of melphalan, cyclophosphamide, VP-16, and adriamycin. The effect was evaluated by the specific growth delay (SGD), computed from transformed Gompertz growth curves. The results showed great variation in the inter-tumor as well as in the intra-tumor response with SGD-values ranging from 1 to 12 and from 0 to 15, respectively. The wide variation in response stresses the need for representative tumor panels in preclinical testing of new drugs and analogs.

Human Carcinoma Cell Lines Xenografted in Athymic Mice: Biological and Antigenic Characterization of an Intra-Abdominal Model

E. TAGLIABUE, S. CANEVARI, S. MENARD, D. MEZZANZANICA, R. ORLANDI, M. RIPAMONTI, and M. I. COLNAGHI

Division of Experimental Oncology E, Istituto Nazionale Tumori, Via Venezian 1, 20133 Milano, Italy

In the perspective of *in vivo* clinical application of monoclonal antibodies (MoAbs) directed against human ovarian carcinoma a preclinical *in vivo* model was developed using athymic mice. Three human carcinoma cell lines (MCF7, HT29 and SW626) were injected into the peritoneal cavity of pristane primed animals and the biologic and antigenic characteristics of the i.p. grown tumors were studied. Tumor take was observed in 85% of the injected animals and palpable nodules became apparent in 83% of the cases 14–27 days after injection. Examination at autopsy of the peritoneal cavity revealed tumor masses varying in size from 0.2 to 0.5 cm in diameter and tumor sheets implanted in the diaphragm. The liver, bowel mesentery and pancreas were also invaded. Ascitic fluid formation was rare and no animal developed vivo tumors outside the peritoneal cavity. To determine whether the *in vivo* tumors retained the same antigenic characteristics as the *in vitro* cell lines, 4 MoAbs directed against ovarian carcinoma associated antigens and 2 different experimental approaches (immunofluorescence and immunoblotting) were used. When immunofluorescence was used the MCF7 line showed the same antigenic pattern both *in vivo* and *in vitro,* whereas the HT29 and SW626 cell lines grown *in vivo* showed a drastic reduction and in some cases the disappearance of the antigens detected by 2 of the 4 MoAbs tested. When immunoblotting was used there was no evident change in the molecular form of the antigens.

This work was partially supported by grants from the Italian National Research Council, contract Nrs 85.02067.44 and 85.01499.57.

Squamous Cell Carcinoma Xenografts as a Model for the Study of Drug-Induced Cell Kinetic Changes in Head and Neck Cancer

J. WENNERBERG

ENT Department, University Hospital, 221 85 Lund, Sweden

During the last 15 years human tumours xenografted to nude mice have been used extensively for the testing of chemotherapeutic drugs. There are, however, only a few studies concerned with drug induced cell kinetic changes in this type of tumour.

We have evaluated xenografted squamous cell carcinoma (SCC) of the head and neck as a model for cell kinetic changes. At transplantation there was a predilection for growth of tumours with a more malignant phenotype according to the histopathological malignancy grading system (MGS) used to classify the tumours. The overall rate of established tumour lines was 16/85 (19%) however, only 4/56 (7%) of tumours with a MGS score of between 8–12 points established tumour lines in contrast to 12/29 (41%) of tumours with a MGS score of 13–17 points. In some attempted transplantations with 2 stem lines in the patient tumours, one diploid and one aneuploid, only the aneuploid cell lines could be recovered from the growing xenografts. Thus, there is a selection of phenotypically malignant tumours and the panel of tumours growing in nude mice are not representative of the panorama of tumours *in situ.* They are, however, probably representative of the tumours that give therapeutical problems.

During the first 3–5 passages in nude mice there is an adaptation to the new environment with an increase in take rate and a decrease in tumour volume doubling time towards a constant value, individual for different tumour lines. These adaptive changes are most likely to depend on changes in cell loss factors, while the cell cycle time (T_c) essentially remains unchanged and in the same order as in human SCC *in situ*, i.e., 1–3 days.

Tumours of established SCC lines exhibit growth curves that can be represented by the Gompertz curve. Furthermore, the variables of the Gompertzian function do not differ between the original tumour and the corresponding xenograft, although the tumour transplants exhibit 10 times faster initial growth and 20 times faster growth retardation than the original tumours in man.

During the course of growth of established tumours there is, with increasing age, a small but significant accumulation of cells in GO/1-phase and a parallel decrease of the fraction of cells in S-phase. These changes are too small, however, to explain the profound changes in tumour volume doubling time (DT) seen during the course of growth. Changes in cell loss factor and growth fraction are most likely the major determinants for the changes in DT.

When squamous cell carcinoma xenografts are used for chemotherapeutic and cell kinetic studies it is of utmost importance to be aware of the changes in tumour growth parameters described above. In the planning of experiments one has to define the growth pattern of the tumours used and to take into consideration the changes of the growth parameters during the course of growth. Otherwise the interpretation of the experiments may either prove be false, or important but small changes in e.g. cell cycle parameters may be overlooked.

With respect to these findings we have used xenografted SCC of the head and neck to study cell cycle and DNA synthesis perturbations induced by i.a. Cisplatin, Bleomycin, Mitomycin C and 5-Fluorouracil. The changes induced differed between the drugs. However, they have in common that cell cycle perturbations are seen very early after drug administration, usually within 24–58 hours, and that these changes precede changes in tumour volume growth.

By this type of stepwise evaluation of single drugs in the nude mouse system it is possible to gain knowledge of the course of induced cell kinetic changes over time. On the basis of such knowledge, with respect to known differences between xenografts and tumours *in situ,* it is then possible to use the model to test different schedules based on cell kinetics and select the most optimal variants for further clinical consideration.

MIX
Papier aus verantwortungsvollen Quellen
Paper from responsible sources
FSC® C105338

If you have any concerns about our products,
you can contact us on
ProductSafety@springernature.com

In case Publisher is established outside the EU,
the EU authorized representative is:
Springer Nature Customer Service Center GmbH
Europaplatz 3, 69115 Heidelberg, Germany

Printed by Libri Plureos GmbH
in Hamburg, Germany